# Fisheries in irrigation systems of arid Asia

FAO
FISHERIES
TECHNICAL
PAPER
**430**

Edited by
**Tomi Petr**
Toowoomba, Queensland
Australia

**2007**
**Published by Arrangement with the**
**Food and Agriculture Organization of the United Nations**
***by***
**DAYA PUBLISHING HOUSE**
**Delhi - 110 035**
**for Sale and Distribution only in South-East Asia**

First Indian Edition 2007

ISBN 81-7035-495-1

*Published by* : **Daya Publishing House**
1123/74, Deva Ram Park
Tri Nagar, Delhi - 110 035
Phone: 27383999
Fax: (011) 23244987
e-mail : dayabooks@vsnl.com
website : www.dayabooks.com

*Showroom* : 4760-61/23, Ansari Road, Darya Ganj,
New Delhi - 110 002
Phone: 23245578, 23244987

*Printed at* : **Salasar Imaging Systems**
Delhi - 110 35

PRINTED IN INDIA

## PREPARATION OF THIS DOCUMENT

This document contains twelve papers presented at the FAO Expert Consultation on the Use of Irrigation Systems for Sustainable Fish Production in Arid Countries of Asia, Almaty, Kazakhstan, 25–29 September 2001. The meeting of experts was organized with the close cooperation of the Kazakhstan Scientific Research Institute of Fisheries, Almaty. The Report of the Expert Consultation is published as FAO Fisheries Report No. 679.

## ACKNOWLEDGEMENTS

Thanks are expressed to the Director of the Kazakhstan Scientific Research Institute of Fisheries, Dr Khismet Ismukhanov and his staff for preparing the venue, for the careful preparatory work and for support during the Expert Consultation.

**Distribution:**
Authors and participants
FAO Regional and Subregional Fishery Officers
Directors of Fisheries
FAO Fisheries Department

**Petr, T. (ed.)**
**Fisheries in irrigation systems of arid Asia.**
*FAO Fisheries Technical Paper*. No. 430. Rome, FAO. 2003. 150p.

## ABSTRACT

**This Fisheries Technical Paper is a companion to the Report of the FAO Expert Consultation on the Use of Irrigation Systems for Sustainable Fish Production in Arid Countries of Asia (FAO Fisheries Report No. 679). The consultation was held at Almaty, Kazakhstan, from 25 to 29 September 2001.**

**This document brings together twelve papers that review the present use of irrigation systems for fisheries in the countries of the arid belt of Asia, from Turkey to China. The individual papers deal with the following countries and areas: Xinjiang Uygur Autonomous Region (China), India, the Islamic Republic of Iran, Kazakhstan, Kyrgyzstan, Mongolia, Pakistan, Turkey and Uzbekistan. While some countries in the arid belt are well advanced in the field of fish production from waterbodies of irrigation systems, others are experiencing major difficulties arising from recent changes in their political and economic systems, particularly Mongolia and the countries of Central Asia. Such countries require major assistance. The document includes a summary of recommendations and proposals for further action, as formulated by the expert consultation.**

# FOREWORD

The FAO Expert Consultation on the Use of Irrigation Systems for Sustainable Fish Production in Arid Countries of Asia was held in Almaty, Kazakhstan, from 25 to 29 September 2001. The meeting of Experts was organized with the close cooperation of the Kazakhstan Scientific Research Institute of Fisheries (Dr Khismet Ismukhanov, Director). The Expert Consultation was attended by 19 participants from 10 countries: China, India, the Islamic Republic of Iran, Kazakhstan, Kyrgyzstan, Mongolia, Pakistan, the Syrian Arab Republic, Turkey and Uzbekistan. Among the participants were two irrigation experts who made possible close interaction on common problems, especially in areas where fisheries interests in water resources have been in conflict with demands from irrigated agriculture.

The major objectives of the FAO Expert Consultation were to review the current achievements and identify constraints in managing fish stocks in irrigation systems of arid countries of Asia, and to assess the potential and develop strategies for further improvement of inland fisheries in water bodies serving irrigation and arising from it.

The arid belt of Asia extends from the Mediterranean to the Pacific, including the following countries: Turkey, the Syrian Arab Republic, Iraq, Iran, southern Russia, countries of Central Asia, i.e. Turkmenistan, Uzbekistan, Kazakhstan, Kyrgyzstan, Tajikistan, and also Pakistan, India, Mongolia and China. In all these countries, with the exception of Mongolia, much of the food production comes from irrigated agriculture and all have inland water resources with harvestable fish stocks.

While FAO and other organizations have been actively promoting the integration of fish production into irrigated small-scale farming systems in high rainfall areas of Asia and Africa, small-scale systems are less common in countries of the arid zone of Asia, especially north and west of India and Pakistan. In these countries large irrigation systems predominate, with water distributed to large command areas for production of food crops and cotton. Such systems include reservoirs, distribution canals, drainage canals, and storages of residual water. Although many reservoirs have been used for fish production with varying success, other types of water bodies have been utilized much less or not at all.

Fishery and agriculture experts from the ten countries who met in Almaty, presented the status of irrigated agriculture and the use for fish production of waterbodies serving irrigation. The agriculture irrigation specialists present allowed for an intersectoral exchange of ideas. This also contributed to a better understanding by agriculture engineers of some problems facing fisheries managers when dealing with fish production in irrigation waterbodies.

Xinjiang Uyghur Autonomous Region, situated in northwestern **China**, is arid, with very low precipitation. Rivers and lakes are fed predominantly by snow and ice melt waters draining from high mountain ranges. Capture fishery, which rapidly developed after World War II, exhausted the available fish stocks and the situation worsened when water from rivers started to be diverted for irrigation to increase crop production for the fast growing population. Fish introductions and species transfers were the first management measures aimed at increasing fish yields in reservoirs and lakes. More recently, fisheries management has concentrated on aquaculture development, which now produces 84.6 percent (47 760 tonnes) of the total fish production in the landlocked Xinjiang Region.

In **India,** crop production from irrigated agriculture has kept pace with the growth of the population: from 51 million tonnes in 1951 (population 361 million) to 200 million tonnes in 1999 (population 1 billion). This can be attributed primarily to the implementation of irrigation projects. India has the largest irrigated area in the world. Man-made reservoirs harnessing water for irrigation, flood control and hydropower have a combined area of around 40 000 km$^2$, ranging from less than 1 km$^2$ to 740 km$^2$. Fish production from Indian reservoirs has been steadily rising and contributes a significant percentage to the total inland fish production. India has shown promising results during pilot studies of fish production in irrigation canals, and has identified waterbodies resulting from seepage and drainage waters as having good potential for pisciculture.

In **Iran** irrigated lands cover 8 million hectares (ha) and those under dry farming cover 6.5 million ha, but the latter category has been subject to serious damage resulting from the recent drought. Limited water resources and the widespread use of unscientific and traditional irrigation systems, which result in up to three times higher water consumption than actually needed for production of crops, are the major constraints facing agriculture in Iran. There are about 3 500 inland waterbodies. Of these about 730, covering a total area of 508 000 ha, are used for fisheries. During 1995–1999 the annual fish production from these waterbodies averaged 30 000 tonnes, employing over 130 000 fishers. It has been recognized that to keep the fisheries sustainable, regular stocking is necessary. About 600 000 ha of land in Iran is allocated to rice farming. It is planned to further develop integrated rice-carp farming and trout farming during the post-harvest period. Trout farming would follow when the water temperature does not exceed 20 °C. Farming in rice fields in 1999 resulted in a harvest of 126 tonnes of fish. The side benefits of this type of culture are fertilization of fields by fish and control of the rice stem borer by carp feeding on it.

In **Kazakhstan** about 70 percent of water is used for irrigated agriculture. Reservoirs in southern and western Kazakhstan serve mainly irrigation, while those situated in the northern half of the country are mostly multipurpose. By the beginning of the 1990s, 2.3 million ha of land were irrigated, plus 1 million ha of irrigated pastures in the deltaic region adjacent to the Aral Sea. Apart from the large number of reservoirs, there are 96 400 km of irrigation canals and 14 900 km of drainage canals. The change from a centrally planned to a market-oriented economy has had a negative impact on land use, resulting in a reduction of irrigated land by about 1 million ha. The system of irrigation canals has deteriorated in a number of places, and no new irrigation systems have been constructed. There has been a drastic reduction in the officially reported fish catches from reservoirs and other waterbodies, in spite of the government maintaining its policy of regular stocking of reservoirs with hatchery-produced fingerlings, as fish hatcheries are still government-owned.

Since much of **Kyrgyzstan** is covered by high mountains, the predominant type of agriculture is livestock production on mountain pastures. Consequently, this Central Asian country has the least developed fishery of all countries in the subregion.

At present, **Mongolia** has no fish production in irrigation systems. In the 1990s, 90 000 ha of land were irrigated, but there are no major reservoirs and irrigation canals which could be utilised for fish production.

In **Pakistan**, 14 million ha are irrigated. There are three major reservoirs, plus 19 barrages, 12 interlink canals and 43 independent irrigation command areas. The total length of main canals is 58 500 km. About 79 percent of the total wheat crop comes from irrigated fields. None of the major dams include fish passes, but some barrages have fish passes which, however, are largely non-functional. Most pumping stations have no fish protection devices and where these are present, they do not function well. In the year 2000, 144 000 tonnes of fish were captured from rivers, reservoirs, natural lakes and irrigation canals; 36 000 tonnes were produced in aquaculture, and about 75 percent of these were produced in fish ponds fed by irrigation canals. There is potential for enhancing fish production in irrigation and multipurpose reservoirs, and also in the at present largely unexploited brackish waterbodies formed from drainage water. Abandoned irrigation canals are used for fish production and managed as fish ponds. Flood control compartments ranging from 10 to 5 000 ha are naturally stocked with fish during floods, but drained a few months later, which does not allow the fish to grow to market size. With proper management, these structures could be better utilized, perhaps for the production of stocking material.

**Turkey** has experienced intensive dam construction activity aimed at tapping as much of its water resources as possible, mainly for hydroelectricity and irrigation. Turkey has 8.5 million ha of economically irrigable lands and 151 dams which have been constructed to store irrigation water. The South Eastern Anatolian Project (GAP) is a multipurpose project which comprises 22 dams and is expected to double the current hydropower and agriculture production of Turkey. While major fish produced in waterbodies of irrigation systems are common carp and trout, Turkey intends to initiate breeding, stocking and production of other indigenous fish species.

**Uzbekistan** uses about 85 percent of the total water runoff for irrigated agriculture, producing mainly cotton, rice and wheat. Water for irrigation is taken from the middle courses of rivers, and drainage water is returned to the rivers further downstream or collected in depressions in lakes without outflow. The total length of irrigation canals is 150 000 km, and drainage canals 100 000 km. Fisheries is practised in numerous bodies of irrigation systems, including waterbodies established from residual water. Much of the fisheries have been privatized, with the former state fishing companies now united under one private management enterprise. Today the major fish production comes from aquaculture in fish ponds established alongside irrigation systems. Twelve of the existing fish farms use irrigation water, and eight use drainage water with salinities of up to 6 g/litre. After a sharp decline in fish production in the early 1990s, fish production in 2000 in all types of freshwater bodies of Uzbekistan reached 9 200 tonnes, of which 6 200 tonnes came from aquaculture and the rest from capture fisheries. However, the fishery potential of irrigation reservoirs is grossly underutilized. The current fish food consumption in Uzbekistan is only 1 kg/person/year.

Fishery management of irrigation reservoirs is best developed in India and Pakistan, where for many years irrigation systems have been a valuable source of fish. Enhancement of fish stocks by stocking, practiced in most countries of the region, has resulted in steady increases in fish production, providing a growing supply to the markets. In countries of Central Asia and in Mongolia fishery management in irrigation and other types of reservoirs has been declining since the dissolution of the Soviet Union. While fish stocks in reservoirs in the countries of Central Asia are presently underutilized, in the Xinjiang Uyghur Autonomous Region of China and in the Syrian Arab Republic fish stocks in reservoirs appear to be overexploited.

Adequate water supply during the critical reproduction and growing periods is a precondition for maintaining fish stocks in irrigation reservoirs. However, due to the priority given to irrigation and often also hydropower production, a satisfactory water supply for fish cannot always be safeguarded. Then other fish stock management methods may have to be applied. The most commonly used method is regular stocking of reservoirs with hatchery-produced fingerlings. Information available for Indian reservoirs indicates that proper stocking of reservoirs significantly contributes to incremental fish production.

Only a few countries use drainage/waste water canals for fish production. Drainage waters have been used for fish production in some parts of India, and information is available from a pilot study in the Syrian Arab Republic. The issue of regular maintenance of drainage canals must be addressed.

Representation of fishery specialists in planning, management and decision-making processes would mean that the interests of fisheries are presented and discussed at a high level. This has not always been the case. Such an approach would contribute to a better understanding of fishery problems faced by other users of the same water resource. As irrigation demands determine the quantities and timing of water discharges, close collaboration is imperative in decisions regarding water allocation and releases. With the ever-increasing demand for water resources, more care must be taken to satisfy all uses, especially for irrigation, hydropower production and fisheries. Reducing water demand for agriculture would have a positive impact on fish stocks and fisheries, especially where current large seasonal agricultural demands cause a major drop in water level damaging the spawning and nursery areas.

On the basis of the papers presented and the follow up discussions, the Expert Consultation put forward a number of recommendations which are published in the FAO Fisheries Report No. 679.

# CONTENTS

# FISH STOCKS AND FISHERIES IN IRRIGATION SYSTEMS IN ARID ASIA

**by**

**T. Petr**
**27 McLeod Street, Toowoomba, Qld 4350, Australia**

## ABSTRACT

The arid zone of Asia extends from the Mediterranean to the Pacific, including the following countries: Turkey, the Syrian Arab Republic, Iraq, Iran, the Russian Federation, Afghanistan, Mongolia, Pakistan, India, China, and countries of Central Asia, i.e. Turkmenistan, Uzbekistan, Kazakhstan, Kyrgyzstan and Tajikistan. In all of them most food crops are produced using irrigated agriculture. In Central Asia, including southern Kazakhstan, over 80 percent of the total water use is for irrigated agriculture. In Pakistan 78 percent of the arable land depends on irrigation as compared with 100 percent in Egypt, 33 percent in all Asia, 21 percent in Near East and Africa, 8.5 percent in Latin America and 2.7 percent in sub-Saharan Africa. In Central Asia, at the end of the 1980s there were over 40 large reservoirs already constructed, with their primary purpose being storage of water for irrigation, closely followed, in some cases, by hydropower production. Fisheries managers have had a number of options for fishery development in the irrigation systems, including reservoirs, major irrigation and drainage canals, and waterbodies established from residual irrigation water. Most options have been a response to the impacts caused by the manipulation of water resource for purposes other than fisheries. Stocks have been enhanced by introductions of species with known preference for such waterbodies and by regular stocking of seed produced in hatcheries. In slow flowing large canals and side storage reservoirs, such as those alongside the Karakum canal in Turkmenistan, fish with pelagic eggs, such as Aral barbel, razor fish, the introduced Chinese carps and white Amur bream (*Parabramis pekinensis*) have been doing well. Some of the drainage collecting depressions have been utilized; the Aral Sea catchment has 2 341 of these depressions, covering 7 066 km² surface area. However, they are characterized by elevated salinities and high concentrations of agrochemicals, the latter being a major obstacle for using fish for human consumption. The fisheries management in waterbodies of irrigation systems of the arid countries in Central Asia has been facing not only major environmental problems, but also those arising from the transition to market economy. Co-management and community-based management of irrigation waterbodies is being considered among the options which could be applied under the new privatization policy. However, in the newly independent countries of Central Asia, the major problem is now lack of funds for urgently needed rehabilitation of some of the existing structures, as well as for the development of small-scale fish enterprises as an alternative to the still prevailing large fish production and fishery systems inherited from the Soviet era, which can no longer be maintained.

## 1. INTRODUCTION

In developing countries approximately 80 percent of water used is for irrigation. It is estimated that in the year 2010, 45 percent of the total global food production will come from irrigated lands. To develop irrigation on such a large scale, many engineering works are needed. At the beginning of the 1990s there were over 36 000 dams in the world (Pircher, 1993), with many of the resulting storages serving irrigation. Some rivers now have a cascade of reservoirs, with little or no free flowing water in between them. Irrigation systems, which consist of dams storing water in reservoirs, of main irrigation canals and their distributory net, drainage canals, and often drainage water storages, offer a diversity of waterbodies for fish production. On the other hand, such systems exercise a negative impact on fish stocks of the original riverine habitats, which are no longer available to them. The impact of damming rivers on fish stocks is well-known: in reservoirs the number of riverine fish species diminishes and they are replaced by fish species with a preference for standing waters, subject to their presence in the catchment. The period of water permanency in irrigation canals is often a limiting factor on their use by fish. Residual waterbodies have a high evaporation rate under desert and semi-desert conditions, and fish face elevated salinity levels, as well as high concentrations of agrochemicals which may be used in irrigated crop production. While all this represents formidable obstacles for the fish, careful management of some irrigation waterbodies is capable of replacing the losses in fish production due to damming rivers. It is the objective of the Expert Consultation to discuss the fishery management practices, constraints imposed on them by irrigation, and what can be done in the region for enhancing fisheries in irrigation systems.

## 2. SITUATION IN THE ARID ZONE OF ASIA

The arid zone of Asia extends from the Mediterranean to the Pacific, including the following countries: Turkey, the Syrian Arab Republic, Iraq, Iran, the Russian Federation, Afghanistan, countries of Central Asia, i.e. Turkmenistan, Uzbekistan, Kazakhstan, Kyrgyzstan and Tajikistan, Mongolia, Pakistan, India and China. In all of them most food crops are produced using irrigated agriculture. All countries in the arid zone tap their inland water resources for irrigated agriculture, and many of the waterbodies have harvestable fish stocks. In Central Asia, over 80 percent of the total water use is for irrigated agriculture. In Pakistan 78 percent of the arable land depends on irrigation as compared with 100 percent in Egypt, 33 percent in all Asia, 21 percent in Near East and Africa, 8.5 percent in Latin America and 2.7 percent in sub-Saharan Africa. In 1987, in 93 developing countries of the world, a total of 164.7 million ha of land was irrigated. This was expected to increase to 220 million ha by year 2000 (FAO, 1993). The irrigation demand ranges from the extreme of the total diversion of water of some rivers for irrigation, to partial diversions and use, with the consequence of various degrees of flow diminution downstream of water abstraction.

In Central Asia, at the end of the 1980s there were over 40 reservoirs constructed or under construction (Nikolaenko, 1988). In the basin of the Syr-Darya Salikhov and Kamilov (1995) counted 22 reservoirs, covering a total of 1 854 km², and 17 reservoirs (1 463 km²) in the basin of the Amu-Darya. In Uzbekistan all four major rivers, Syr-Darya, Zarafshan, Kashka-Darya and Amu-Darya, and their tributaries have been regulated and have storages for irrigation (Kamilov and Urchinov, 1995). Their major purpose is to keep sufficient amount of water of required quality for irrigation use. An example of the increasing demand for irrigation water in Turkmenistan was given by Kostyukovsky (1992): while in 1900, with a total population of 1 444 000, there were 5 530 km² of irrigated land, requiring 3.68 km³ of water, by 1986 (population 5 446 000) this rose to 17 830 km², which required 16.50 km³ of water. Micklin (1991) produced a map of the major irrigation zones in Kazakhstan and Central Asia (Fig. 1).

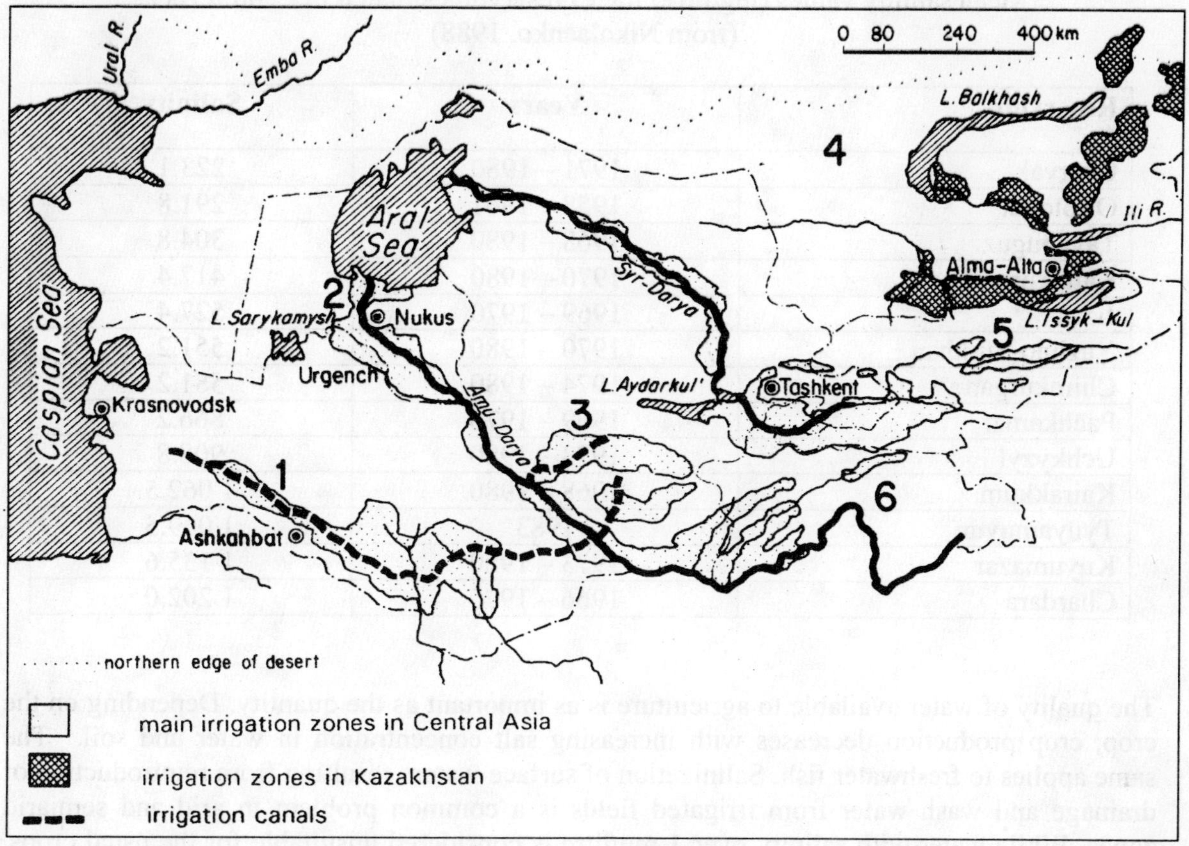


**Fig. 1** Major irrigated areas in Central Asia and Kazakhstan (Micklin, 1991)
1. Turkmenistan; 2. Karakalpakstan; 3. Uzbekistan; 4. Kazakhstan; 5. Kyrgyzstan; 6. Tajikistan

Under arid conditions of Central Asia the average evaporation rate is two-fold that of precipitation. This leads to salinization of water and soils. The long-term average salinities of reservoirs in the montane and foothill zone above 500 m altitude range from 223 to 527 mg/litre, with maximum salinities reached during winter and spring, before floods. In lowland reservoirs the salt concentration ranges from 550 to 1 200 mg/litre, with a maximum of 1 700 mg/litre reached during the autumn-winter period (Nikolaenko, 1988). Nikolaenko (op.cit.) also compiled information on average salinities of 13 reservoirs of Central Asia (Table 1). It shows that by the mid-1980s in four out of 13 reservoirs salinities exceeded the value of 1 g/litre. Water quality in the reservoirs of Central Asia started deteriorating in 1974 when drainage waters, with a high concentration of sulphates, chlorides, manganese and sodium, were diverted back into rivers.

**Table 1**
Mean salinity values (mg/litre) for 13 reservoirs situated in Central Asia.
(from Nikolaenko, 1988)

| Reservoir | Years | Salinity |
|---|---|---|
| Charvak | 1971 – 1980 | 223.1 |
| Ortotokai | 1958 – 1961 | 291.8 |
| Tayabuguz | 1968 – 1980 | 304.8 |
| Kattakurgan | 1970 – 1980 | 417.4 |
| Jizak | 1969 – 1970 | 527.4 |
| Yuzhno-Surkhan | 1970 – 1980 | 551.2 |
| Chimkurgan | 1974 – 1980 | 581.2 |
| Pachkamar | 1969 – 1976 | 866.2 |
| Uchkyzyl | 1973 – 1980 | 908.8 |
| Kairakkum | 1968 – 1980 | 1 062.5 |
| Tyuyamuyun | 1983 | 1 069.5 |
| Kuyumazar | 1973 – 1980 | 1 135.6 |
| Chardara | 1966 – 1976 | 1 202.0 |

The quality of water available to agriculture is as important as the quantity. Depending on the crop, crop production decreases with increasing salt concentration in water and soil. The same applies to freshwater fish. Salinization of surface waters resulting from reintroduction of drainage and wash water from irrigated fields is a common problem in arid and semiarid zones. While water with salinity over 1 mg/litre is considered unsuitable for the usual crops, less is known about water salinity levels harmful to especially the young of the native fish of arid and semiarid climates. Agricultural practices also affect nutrient concentrations in surface and groundwaters. In surface waters, this may result in a ready availability of nutrients to aquatic plants, leading to excessive growth of plants in irrigation and drainage canals. Antagonistic interactions between the agriculture and fisheries sectors arise from the application of pesticides and herbicides which can be harmful to aquatic living organisms. Agrochemicals used against pests or for defoliation, such as in cotton production, contribute to serious water quality problems and represent a hazard to fish and the end consumers – birds and man. Where drainage and wash waters are diverted in desert depressions without an outflow, such as Lake Sarykamysh in Uzbekistan, or into swamps, agrochemical concentrations may gradually reach unacceptable levels making the fish unsuitable for human consumption.

## 3. FISH STOCKS AND FISHERIES IN IRRIGATION SYSTEMS OF THE ARID ZONE OF ASIA

Irrigation systems require a novel approach to fishery management. Where large reservoirs have formed, riverine fish may not find the new habitat suitable for all periods of their life cycle, and their number will gradually decline, with some species disappearing completely. While a new management approach is required, some of the knowledge on reservoir fish and fisheries can be adapted from similar situations where a river was dammed for hydropower electricity production. Indeed, reservoir fisheries in some parts of arid Asia have been well advanced, for example in some countries of Central Asia and in India. Stocks have been enhanced by introductions of species with known preference for such waterbodies. Elsewhere, or in addition to introductions, reservoirs have been regularly stocked with fingerlings produced in hatcheries. This has resulted in a sustainable production of fish from reservoirs such as Chashma in Pakistan and Kapshagay in Kazakhstan. Irrigation and hydropower reservoirs often experience a large and untimely drawdown which prevents the desired fish species from finding suitable spawning grounds. The inherent productivity of the reservoir is also important, with reservoirs with a large throughflow rate usually poor in fish food such as phyto- and zooplankton. Introductions and regular stocking of exotic fish, while increasing fish production, may lead to a change in fish species composition. Often the usually highly adaptable introduced species gradually replace the indigenous fish. While there may be a decline in indigenous fish species, introductions usually result in an increase in species diversity. But not always the final result is a substantial gain in fish production as known for the Kapshagay reservoir in Kazakhstan, where to maintain fish stocks for sustainable fisheries it is necessary to regularly stock this reservoir. As introductions may not result in a permanent establishment of all of the introduced species, to prevent a decline in fish catches fisheries managers may also pursue the strategy of bringing in additional exotic fish species to sustain the fish production and fishery. Eventually, such reservoirs may become inhabited largely by the exotics.

By blocking the migratory path of fish, dams have a major impact on fish species which require suitable spawning and/or nursery and feeding grounds. Dams on the Amu-Darya and Syr-Darya have blocked the migratory path of fish, such as Aral barbel (*Barbus brachycephalus*), shovelnose (*Pseudoscaphirhynchus kaufmanni*), sturgeon (*Acipenser nudiventris*), Aral trout (*Salmo trutta aralensis*), and pike asp (*Aspiolucius esocinus*) now threatened with extinction (Pavlovskaya, 1995). In 1974 a fishway opened on the Takhiatash dam on the Amu-Darya, about 200 km upstream from the Aral Sea, enabling a small scale upstream migration. Dams on the lower Indus and the Ganges have stopped the migration of hilsa (*Hilsa* spp). The large consumptive irrigation use of water from the Amu-Darya and Syr-Darya has impacted not only the fish stocks and fisheries in the terminal Aral Sea, but it also led to a decline in fish stocks and fisheries in the deltas of these rivers. (See also paper by Kamilov, this volume.) In northwestern China the total diversion of water for irrigation from several rivers has resulted in the disappearance of fish and fisheries. Stocks of the slow growing naked carp (*Gymnocypris przewalskii*) in the saline Lake Qinghai in the Qinghai Province of China have been hard hit by the combined impact of overfishing, blockage of the inflowing rivers by weirs preventing their upstream migration into rivers for spawning, and water diversion for irrigation (Walker and Yang, 1999).

Damming of the Ili River in Kazakhstan has had a negative impact on fish stocks and fisheries of the terminal Lake Balkhash. Retention of water in the Kapshagay reservoir on the Ili, and its partial diversion for irrigation, led to an increase in the salinity of Lake Balkhash and a consequent decline in fish landings (Petr and Mitrofanov, 1998). The desiccation of wetlands and lakes in the Ili delta resulted in shortages of food and spawning areas for carp and pikeperch, both fish known for their tolerance of waters with elevated salinities. The instability in the Ili River discharges, while caused largely by the highly variable rainfall over the years, has made the future of fish stocks in Lake Balkhash difficult to predict. No doubt, the presence of the Kapshagay dam has further contributed to the problems faced by fisheries. Much has been written about the impact of water diversion for irrigation from the Syr-Darya and Amu-Darya rivers on the fate of the Aral Sea.

# 4. REHABILITATION AND ENHANCEMENT OF FISH STOCKS AND FISHERIES

Fisheries managers have a number of options for fishery development in the irrigation systems. Most options are a response to the impacts caused by the manipulation of water resource for purposes other than fisheries. As it is difficult to predict the weather pattern for each year, irrigation systems are subject not only to rainfall, but in some places also ambient air temperature, which determines the amount of snow- and ice-melt from glaciers, and evaporation rates. Agriculture practices such as selection of crops are also not static, and may change from year to year, and with them the amount of water and timing of the demand required for achieving the best crop. Furthermore, there is the use of agrochemicals, which may differ from year to year. Thus, fisheries managers, while having an overall master plan, may also need contingency plans, as fisheries in irrigation systems are subordinated to other demands for water. In addition, the fishery sector is usually underfunded by the government. The positive aspect of the former centrally planned economies of some countries in the Region was that they had a well-organized research and collection of statistics obtained through regular monitoring of fish stocks, so that the impacts of introductions, stocking, and catches could be evaluated and management strategies adjusted. Centralized statistics provided the longest series of data for a number of irrigation reservoirs situated in Central Asia, and these can now be used for further assessment, evaluation and as examples of the level of efficiency of the applied enhancement measures. They also show the failures resulting from some introductions. The following section provides information on four reservoirs in Central Asia and Kazakhstan, for which long series of data are available. Much of this information is abstracted from Karpova, Petr and Isaev, 1996.

## 4.1 Reservoirs

Chardara reservoir (formerly called Chardarin) is situated at an altitude of 252 m on the Syr-Darya in Kazakhstan, at its border with Uzbekistan. It is an example of the multiple-purpose of many similar reservoirs, in this case hydropower production, storage of irrigation water, municipal and industrial water supply, and fisheries. Unlike reservoirs in the temperate zone of Europe, the USA and Canada, it fills during autumn and winter. From the end of April/beginning of May to September/October the reservoir water is drawn for irrigation, with water level declining by up to 12 m. At the maximum drawdown the reservoir covers only 11 000 ha, as compared to 90 000 ha when full. There are 32 fish species in this reservoir, of which 14 are of economic importance. While initially bream (*Abramis brama*), the wild form of common carp ("sazan", *Cyprinus carpio*), pikeperch (*Stizostedion lucioperca*) and rasorfish (*Pelecus cultratus*) dominated in fish catches, the introduced and stocked common ("European") carp (*Cyprinus carpio*), which appeared for the first time in catches in 1987, reached 67 percent of the total catch in 1991 (Fig. 2). While the mean annual catch for the 12-year period 1980–1991 was 1 712 tonnes, equal to a yield of 19.0 kg/ha (as for the area of the full reservoir), in 1990 the catch was 2 565.9 tonnes, equal to a yield of 28.5 kg/ha. This was the result of the rapid establishment of the introduced common carp. Other fish species continued to appear in fish landings at approximately the same numbers as prior to the appearance of the common carp, thus indicating that its introduction did not negatively affected them. During the 1990s the reservoir fishery underwent major changes. For details see Ismukhanov and Mukhamedzhanov (this volume).

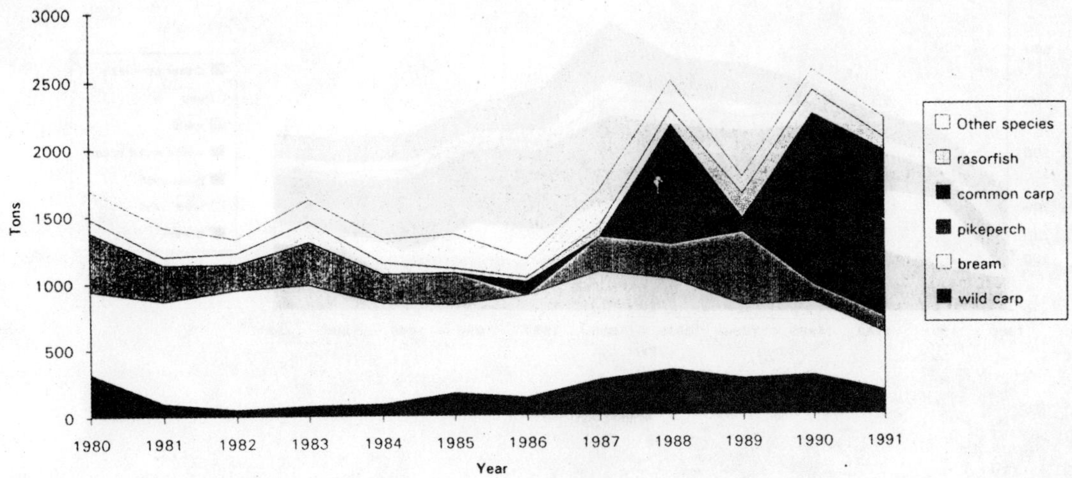


**Fig. 2** Composition of fish catches in Chardara reservoir (Karpova, Petr and Isaev, 1996)

Kairakkum reservoir on the Syr-Darya in Tajikistan is situated at an altitude of 347.5 m, covers 81 300 ha when full, and 26 000 ha at the minimum operating level. It is providing water for irrigation, municipal and industrial demands, and it also has fisheries. During summer the water level in the reservoir sharply declines because much of the water is used for irrigation. This has a negative impact on fish stocks, especially during dry years. The reservoir fish species are the indigenous fish of the Syr-Darya. Bream, pikeperch, and wild carp dominate in fish catches, asp (*Aspius aspius*), wels (*Silurus glanis*), Aral white-eyed bream (*Abramis sapa*) and rasorfish are also species of fishery importance (Fig. 3). The introduced Chinese carps (grass, silver and bighead), which appeared in catches from 1988 onwards, made a minimal contribution to fish landings. The reservoir is not productive, with a mean annual catch of 481.2 tonnes for a 12-year period (1980–1991), which equals to a yield of 4.33 kg/ha. The yield could be perhaps doubled or tripled with sustained stocking, if there were to be a sufficient food base for the stocked fish.

The establishment of fish fauna in Kapshagay reservoir on the Ili River in Kazakhstan has been thoroughly studied by Mitrofanov *et al.* (1988, 1992). The Ili River enters Kazakhstan from China, and downstream of Kapshagay reservoir it ends in the slightly saline endorheic Lake Balkhash (Fig. 4). Kapchagay reservoir covers 184 700 ha when full, which happened for the first time in 1988, since the dam closure in 1969. The reservoir is multipurpose and serves hydroelectricity production, irrigation, transport, as a municipal and industrial water supply, for regulation of water discharges to maintain water quality downstream including in the major canals, and for nature conservation, recreation, and fisheries. Before the closure of the dam, 30 fish species were listed for the river (Petr, 1992). In 1975, 24 fish species were recorded in the reservoir, and only 19 species in the Ili River (Petr and Mitrofanov, 1998). Subsequently the number of fish species in the reservoir declined to 18 species by 1980. Commercial fishery started in 1972, and during the first years following the filling several fish species of economic importance were introduced, such as pikeperch, asp (*Aspius aspius*) and wels from the Ural River, *Chondrostoma nasus* and *Leuciscus lindbergi* from Lake

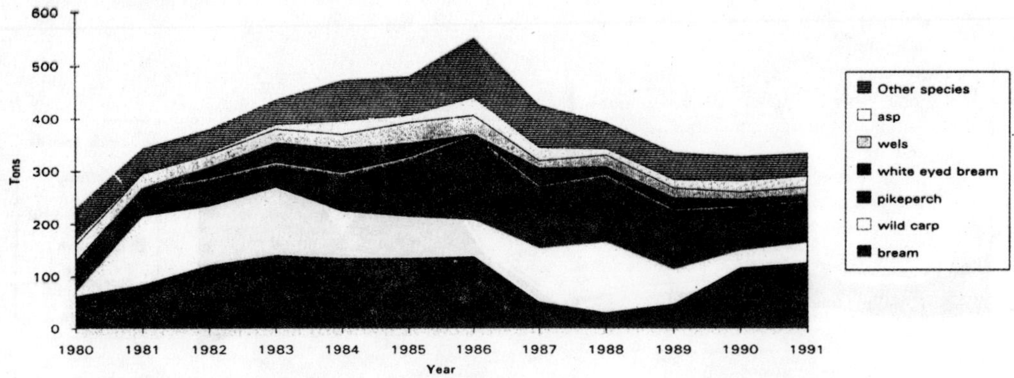


**Fig. 3** Composition of fish catches in Kairakkum reservoir (Karpova, Petr and Isaev, 1996)

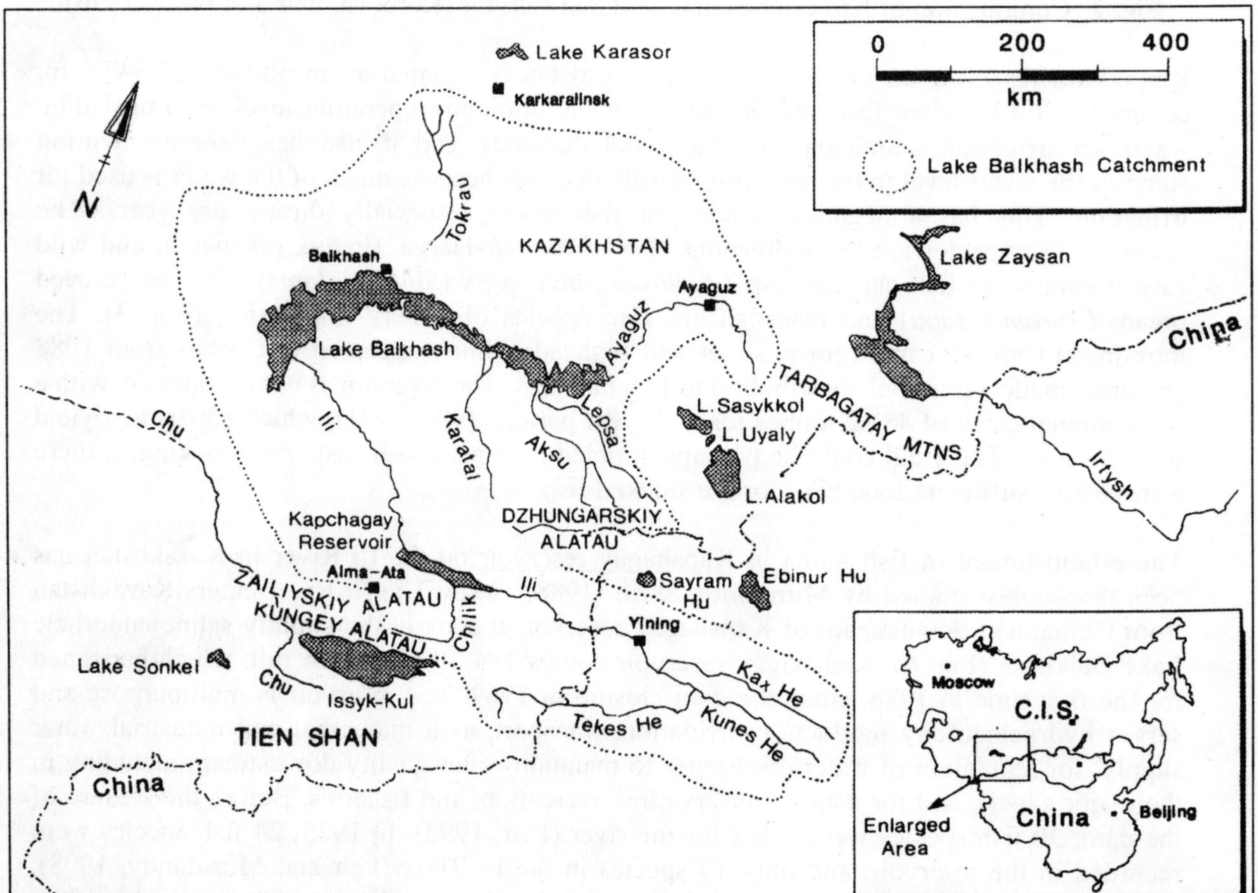


**Fig. 4** Lake Balkhash Basin, Chu River basin, Lake Issyk-Kul (from Petr and Mitrofanov, 1998)

Beilikol, and five species (*Pseudorasbora parva, Pseudaspius leucocephalus, Hemiculter leucisculus, Percottus glehni* and *Rhinogobius similis*) from the Soghua River in Heilongjiang Province in China. The initially dominant common carp was replaced by pikeperch, and this, in turn, by bream. With the water level now being subject to hydropower production and irrigation demands, and not seasons anymore, spawning opportunities for the common carp and some other fish have been greatly reduced. The manipulation of water level has not affected the breeding of bream which has pelagic eggs. Stocks of roach (*Rutilus rutilus*), crucian carp (*Carassius carassius*) and wels have also gradually built up and gained in fisheries importance. Spiny sturgeon (*Acipenser nudiventris*), never common in the river, virtually disappeared. In 1980 the catches were highly dominated by bream, with much less common wels, asp, wild form of the common carp and other species (Fig. 5). During the period 1980–1991 the mean annual fish landings were 1 008.6 tonnes, which equals to a yield of 10.1 kg/ha as per 100 000 ha, which is an estimate of the average maximum water level surface area during the observation period.

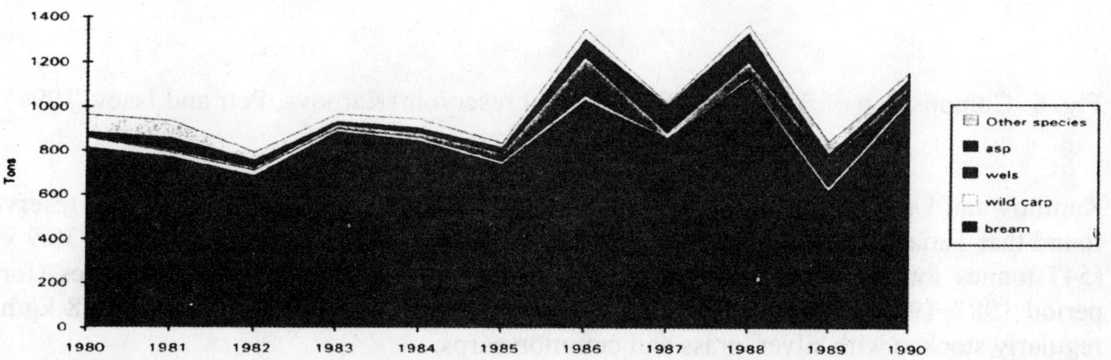


'ig. 5 Composition of fish catches in Kapshagay reservoir ( Karpova, Petr and Isaev, 1996)

<u>Toktogul reservoir</u> in Kyrgyzstan is situated on the Naryn River, in the watershed of the Syr-Darya, at an altitude of 900 m. At a maximum water level it covers 28 400 ha, which is reduced to 15 000 ha at the minimum live storage. The reservoir serves for hydropower production and irrigation. The fish are those of the Naryn River and its tributaries plus a number of introduced species. Initially, in catches dominated the indigenous snow trout (*Schizothorax intermedius*), closely followed by the Amu-Darya trout (*Salmo trutta oxianus*) (Fig. 6). Commercial fishing started in 1978, but there were large differences between annual catches. A certain stabilization was noticed since 1988, when catches reached up to 25 tonnes/year, with the snow trout being the dominant fish in landings. The introduced silver carp (*Hypophthalmichthys molitrix*) and common carp were also a regular, but small component in the catches. More recently the introduced Issyk-Kul (Sevan) trout (*Salmo ischchan gegarkuni*) became an important commercial fish species. For additional information on this reservoir see Djancharov (this volume).

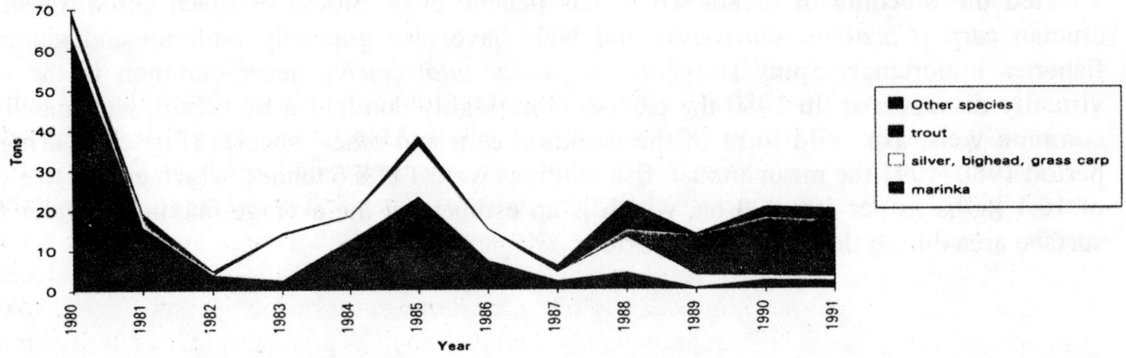


**Fig. 6** Composition of fish catches in Toktogul reservoir (Karpova, Petr and Isaev, 1996)

Kamilov and Urchinov (1995) who compared the fish catches from 12 Uzbekistan reservoirs, found that Tudakul reservoir (17 400 ha when full) was having the highest yield of 30.9 kg/ha (547 tonnes for the whole reservoir), and a mean annual harvest of 453.5 tonnes (for the period 1987–1992). The potential sustainable yield was estimated to be around 78 kg/ha, if regularly stocked with silver, grass and common carps.

Fishery management measures in Uzbekistan reservoirs have included the introduction of several fish species plus their food organisms, and stocking. The results have differed among the reservoirs but usually with the increasing age of the reservoir species diversity has increased, as a result of introductions and of the immigration of fish from rivers and through canals. Fisheries, initially harvesting stocks which formed without human interference, are now paying more attention to the sustainability of stocks. Long-term research on several reservoirs has indicated that the major reasons for the low fish production in some reservoirs are: poor utilization of the natural fish food, poor spawning conditions and nursery habitats, and vacant niches not yet occupied by economically important fish species. In some reservoirs aquatic plants are underutilized, or benthos is utilized by fish species of low value. Fishery management programmes were prepared in the past for the major river systems of Uzbekistan which should lead to increasing fish yields four to six times, but implementation of the programmes has been delayed by the major political and economic changes, which have taken place in the country over the last 10 years. For more details see Kamilov (this volume).

### 4.2 Irrigation canals

The rate of water flow in the major irrigation canals usually ranges from 40 to 300 m³ per second, with a water velocity between 0.4 to 2.0 m per second. In some major canals, when not in use, a minimum flow of 5 m³ may be maintained. Where rivers carry a high sediment load, a sedimentation reservoir may be constructed at the head of the canal. While the average concentration of suspended sediments at the intake of water from the Amu-Darya into Karakum canal is 3.7 g/litre corresponding to a transparency of less than 40 cm, after the water leaves the sedimentation reservoir its transparency is much higher. But as the distance from the reservoir increases, the transparency gets gradually reduced due to the erosion of the canal sides. Apart from current velocity and water transparency, water salinity is also important for fish. In the Amu-Darya irrigation canals salinity ranges from 118 to 1 304 mg/litre (Ergashev, 1989), increasing towards the canal tail reach and final distributaries.

In slow flowing large canals and side storage reservoirs, as present alongside the Karakum canal in Turkmenistan, fish with pelagic eggs, such as Aral barbel, razor fish, the introduced Chinese carps and white Amur bream (*Parabramis pekinensis*) have been doing well. The introduced fish are well established in the Karakum canal and breed there. In a number of irrigation canals grass carp has greatly assisted in controlling aquatic plants. In the Karakum canal, the efficiency of this fish but especially its feeding selectivity has lead to a succession in macrophytes from *Myriophyllum spicatum* to *Ranunculus*, which is considered toxic to grass carp. Total removal of aquatic plants by grazers such as grass carp does not favour some other fish such as common carp, which needs aquatic plants for attachment of eggs, if it is to coexist with grass carp. Grass carp is also known to contribute to the eutrophication of those irrigation canals with dual purpose, i.e. irrigation and as a source of drinking water. Its use for control of aquatic macrophytes therefore has to be carefully planned, especially in desert, where other sources of water are not available. The eutrophicating impact of grass carp can be counteracted by the introduction of silver carp and bighead carp, which will feed on phytoplankton and this may prevent eutrophication of the canal water.

### 4.3 Lakes established from drainage waters

Where drains collecting residual irrigation water and wash water do not re-enter a river and/or a terminal lake, as required when the salinity of the drainage water exceeds 1 mg/litre, the water may be diverted to a depression, where it creates a new water body. Pavlovskaya (1995) estimated the number of drainage collecting depressions in the Aral Sea catchment at 2 341, covering 7 066 km² surface area. More than one third of drainage water collecting depressions and waterbodies are in the Syr-Darya River basin, where the largest is the Arsanay Lake system (4 000 km²). Twenty-four per cent of such waterbodies, with a total area of 52 percent of the total, are concentrated in the Amu-Darya River basin, with the largest, Lake Sarykamysh, exceeding over 3 000 km² in Uzbekistan and Turkmenistan. Kamilov and Urchinov (1995) and Pavlovskaya (1995) provided figures for fish catches for five drainage lakes. In 1992 fish landings from the Arnasay system, which had at that time a water salinity of 10.1 g/litre, were 1 141.4 tonnes. This was less than 40 percent of the maximum of 2 558.4 tonnes harvested in 1990. Only 334.5 tonnes were harvested in 1992 from Lake Sarykamysh, where water salinity has been gradually increasing and by 1992 reached 12.5 g/litre. Five years earlier fishermen harvested from the same lake 1 347.9 tonnes. In 1992, fish landings from three other lakes, established from drainage waters, were: 117.2 tonnes (Dengizkul), 5.3 tonnes (Karakyr), and 4.1 tonnes (Tuzgan). The best information on the impact on fish stocks and fisheries of the changes in the size and chemical character of a water body, established from drainage waters, is that available for Lake Sarykamysh.

Lake Sarykamysh

In Sarykamysh fishery started in 1966, with the major fishery organization located in Turkmenistan. In 22 years of commercial exploitation a total of 21 300 tonnes of fish were harvested. The maximum annual catch of 2 500 tonnes was recorded during the 1981–85 period, with yields ranging from 4 to 8 kg/ha. There was no restriction on fishing and the fishery resource was exploited to its maximum. Uzbek (Karakalpak) fishermen started fishing only in 1980, and stopped in 1988 as fishing became unprofitable due to the large distance between the water body and fish processing plant, and also because of the poor quality of fish (Pavlovskaya, 1995). The Karakalpak fishery area is situated in the north of this lake where the salinity is higher than that in the south. The high salinity prevents fish, such as common carp, Aral barbel and wels, from growing larger. The fishery focused on Aral barbel and common carp and due to the lack of enforcement of regulatory measures the stocks of these two species became overexploited. This was accompanied by an increase in the less valuable pikeperch and razorfish. However, the major problem has been the continuing increase in the lake salinity, which by 1995 was expected to reach 16–18 g/litre.

The Lake Sarykamysh experience has provided a number of lessons on the impact of a rapidly changing aquatic environment on indigenous and introduced fish and their fisheries. Sarykamysh shows the instability of lakes established from drainage water from irrigated agriculture in desert conditions with a high evaporation rate. Instability of especially the limnological environment is inherent in such lakes, which are subject to a gradual increase in salinity. Freshwater fish are stressed as the salinity affects the fertilization and hatching of eggs and retards growth of fish. Studies have shown that the ratio of predatory to prey fish increases, largely due to the increase in the number of pikeperch, which is the most salinity-resistant commercial fish species in the Aral Sea catchment. In Sarykamysh bream, pikeperch and razorfish were the most adaptable and productive fish under the increasing water salinity. Eventually, pikeperch represented 27 percent of the total fish stocks (Sanin and Shaporenko, 1991) but this was followed by their decline (Fig. 7).

The experience with Sarykamysh has shown that waterbodies with increasing salinity need a dynamic environmental and fishery management approach if they are to continue serving fish production. The deteriorating spawning conditions resulting from the increasing salinity may perhaps be compensated by breeding fish in hatcheries and stocking fingerlings. Other species, such as of estuarine character which tolerate large water salinity differences, could be also tested. A major problem is the increasing load of pesticides and other toxic substances applied in agriculture. Alternative solutions should be found for lowering agrochemical application levels and biological control should be introduced where possible (Petr and Mitrofanov, 1998). A certain decrease in the use of agrochemicals has taken place since countries of the Central Asia entered the transition period from centrally planned economy to market-oriented economy.

# 5. DISCUSSION

Interesting information arises from a comparison of the fisheries in the pre-irrigation period in the Chu River catchment (Kyrgyzstan/Kazakhstan) with that of intensive irrigation. During the first period, i.e. in the 1930s, the catches in the River Chu were estimated at 4–5 tonnes in the submontane zone and 500–600 tonnes in the lowland lakes. By the 1960s, the annual

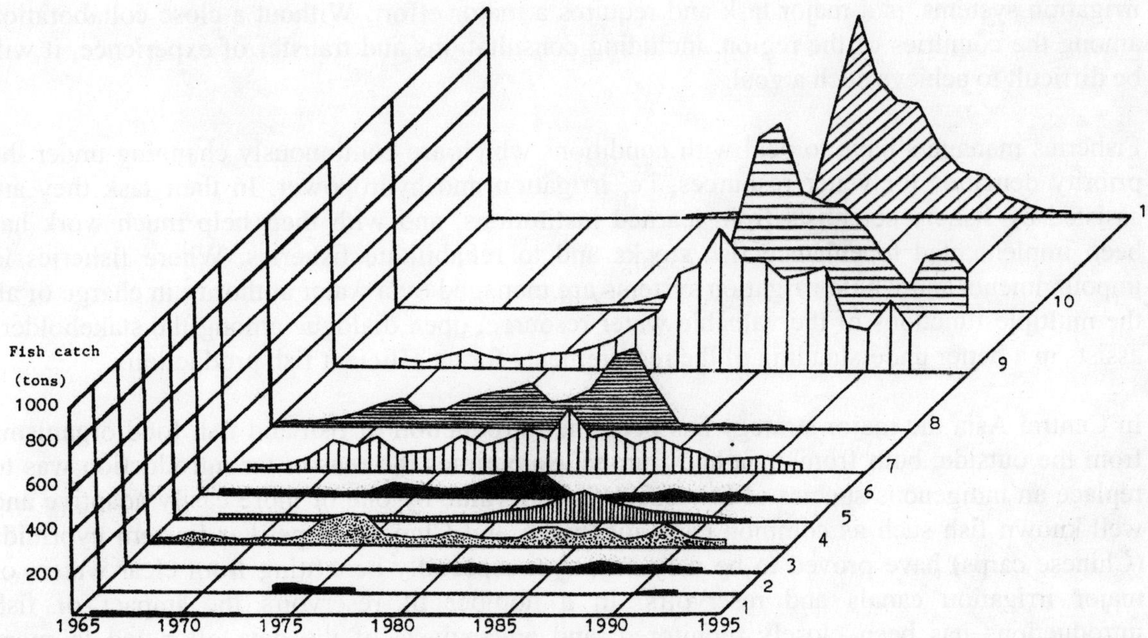


**Fig. 7** The dynamics of the major fish species in catches from Lake Sarykamysh (1965–92). 1. *Esox lucius*; 2. *Chalcalburnus chalcoides aralensis*; 3. *Hypophthalmichthys molitrix*; 4. *Rutilus rutilus aralensis*; 5. *Silurus glanis*; 6. *Aspius aspius iblioides*; 7. *Cyprinus carpio*; 8. *Barbus brachycephalus;* 9. *Abramis brama orientalis*; 10. *Stizostedion lucioperca*; 11. *Pelecus cultratus*. (Petr and Mitrofanov, 1998).

catches in the lakes declined to 250 tonnes and in 1970 to as little as 20 tonnes. When comparing these figures with the volume of water uptake for irrigation, a direct relationship between the diminishing water discharge in the river and fish catch becomes evident. The poor fish catches in 1970 were when there was a maximum uptake of the river water (166 $m^3 s^{-1}$) for irrigation. Between 1930 and 1970, the area of irrigated land doubled from 190 000 to 383 000 ha. By 1970, the total surface area of reservoirs in the montane zone reached 40 $km^2$. Pond fish culture expanded in the submontane zone and commercial fishing developed in Tashutkul reservoir. In 1971, the fish farms produced about 450 tonnes of fish for markets, but afterwards emphasis shifted to the production of stocking material for reservoirs. As a result, in the early 1980s the output of market fish was at the same level as in 1971 and the output of juveniles for stocking was 65 million per year. One can conclude that the total fish output in the watershed of the River Chu has remained the same as it was before the introduction of irrigated farming. However, the cost of producing the same amount of fish became much higher (Petr and Mitrofanov, 1998).

While in arid and semiarid Asia the demand for irrigation water is still rising, the number of dams under construction is much lower than during the second half of the last century, largely because the majority of economically harvestable water resources has been already tamed. It is becoming evident, that irrigation needs to turn to more efficient water use methods. Overexploitation of water resources in arid regions of Central Asia, where water supply depends largely on snow and icemelt in distant mountains, has had a tragic consequences as documented in the desiccation of the Aral Sea and the loss of a major fishery there, which at its peak reached 44 000 tonnes per year. To replace such a quantity of fish through improved production from a diversity of inland waterbodies, almost all of which are now part of irrigation systems, is a major task and requires a major effort. Without a close collaboration among the countries of the region, including consultations and transfer of experience, it will be difficult to achieve such a goal.

Fisheries managers have to deal with conditions which are continuously changing under the priority demands for water resources, i.e. irrigation and hydropower. In their task they are assisted by fishery scientists from learned institutions, and with their help much work has been implemented to enhance fish stocks and to rehabilitate fisheries. Where fisheries in impoundments and whole irrigation systems are managed by a water authority in charge of all the multiple-functions of the valuable water resource, open dialogue among the stakeholders assists in a better understanding of the requirements for an efficient fish production.

In Central Asia the major strategy has been the introduction of fish and fish food organisms from the outside, both from Asia and Europe. Sometimes the reason for introduction was to replace an indigenous species of low or no fishery value by one or more easily adaptive and well-known fish such as common carp, pikeperch and Chinese carps. Far Eastern cyprinids (Chinese carps) have proved to be very adaptive, especially benefiting from clear waters of major irrigation canals and reservoirs. In a number of reservoirs the impact of fish introductions has been closely monitored, and an analysis of the data often led to more introductions. At present, there is probably not a single water body in the region left with only the original indigenous fish fauna, and introduced fish species may represent 50 percent or more in the total catch.

With the rising demand for fish in arid countries of Asia the need for a better management of water resources in irrigation systems is evident. Only in this way the fish production can increase, not only by enhancing the fish stocks in open waters through introductions and stocking, but also through the implementation of fish culture, such as cage culture, raceway culture and blocked-off cove culture. Special attention needs to be paid to fish production in saline waterbodies in the region and ways should be found how to reduce the concentrations of agrochemicals and other pollution of the residual waterbodies. Where swamps are used for such discharges, they should also be assessed for fish production. The possibility of using fish for the control of vectors of water-born diseases should also be taken into account, in some cases combined with the control of excessive growth of aquatic macrophytes as in irrigation canals. Elevating fisheries in irrigation systems from the often subordinated position to an equal partner in discussions and in the decision-making process should appear on the agenda of government deliberations dealing with irrigation systems.

Co-management and community-based management of irrigation waterbodies could be applied under the new privatization policy and market-oriented economies emerging in Central Asia. Maintaining and monitoring the fishery resource by a small group would facilitate exclusion of outsiders, often illegal fishermen. Government policy-makers may consider delegating the management responsibility to such collective or private groups, which then should receive government support through credit, training, scientific and extension assistance.

Outside Central Asia a number of countries in the dry zone faces the problem of how to best develop the wide range of waterbodies for fish production. Turkey has constructed a system of dams on the Euphrates and Tigris rivers that serves irrigation of large areas for agriculture production. The General Directorate of State Hydraulics in Turkey, which is in charge of the ever increasing number of reservoirs, has fostered the strategy of regular stocking of reservoirs with hatchery-produced fingerlings, largely common carp. In Keban and Karakaya reservoirs on the Euphrates the return on stocking has been 13 percent and 19 percent, respectively. During the 10-year period of 1982–1991, 20 million fingerlings were stocked into 115 reservoirs and 68 smaller waterbodies in Turkey, the majority of them having multipurpose function, including irrigation. Since then, at least seven more fish species are hatchery-produced for stocking (Petr, 1998). Pakistan and India have a sophisticated system of barrages, dams and irrigation canals. These countries have also made considerable progress in utilizing the irrigation systems for fish production. The expansion of wetlands in the Indus Valley and Indus delta in Pakistan is a direct result of the development of irrigation and river management. Cross (1994) noted that environmental assessment of water storage development in Pakistan has generally failed to take into account the positive value of the appearance of new wetlands.

Where there is irrigation, there is also seepage. Waterbodies from seepages from reservoirs and canals are not uncommon. Seepage from a dam can be utilized for hatcheries, ponds, or for raceways for production of trout, as for example in Iran and China (Petr, 1998). In Pakistan, a system of ponds has been constructed alongside the main canal diverting water for irrigation from Chashma reservoir on the Indus. Where seepage water accumulates under arid conditions with high evaporation rates saline waterbodies may form. In a suitable climate such "derelict" waterbodies can be used for culturing salt-tolerant species, such as tilapias.

The almost complete exploitation of surface waters in some arid areas of Asia is becoming a major obstacle to developing further areas under extensive furrow-type irrigation, which dominates irrigated agriculture in most of the countries. The central question confronting authorities is how to better utilize the existing water resources for the existing agriculture, and for its expansion. It is understood, that a fundamental change in agricultural use of water is needed, i.e. conversion from largely extensive agriculture, demanding high volume of water, to well-managed intensive agriculture based on low water demand. Shortening the flooding period in rice fields has had some negative impact on rice-fish culture in paddy fields in the Southeast Asia. As rice production is not the major crop in the arid belt of Asia, such a change may not affect the fish and fisheries in higher latitudes, where storages and distribution systems of canals will continue to represent the major waterbodies with fishery potential.

## REFERENCES

Cross, D., 1994. Getting wetter. New Scientist 1936: p. 51.

Ergashev, A.E., 1989. Central Asian irrigation canals and some aspects of their algal flora. Int. Rev. ges. Hydrobiol. 74: 83–94.

FAO, 1993. World Agriculture Toward 2000. FAO, Rome.

Kamilov, G. & Urchinov, Zh.U. 1995. Fish and fisheries in Uzbekistan under the impact of irrigated agriculture. FAO Fisheries Circular. No. 894: 10–41. FAO, Rome.

Karpova, E.I., Petr, T. & Isaev, A.I. 1996. Reservoir fisheries in the countries of the Commonwealth of Independent States. FAO Fisheries Circular. No. 915. FAO, Rome. 131p.

Kostyukovsky, V.L., 1992. Dynamics of Turkmenistan lowlands in conditions of natural and man-induced changes. Vodnye Resursy No. 4: 85–92. (In Russian)

Micklin, P.P., 1991. The water management crisis in Soviet Central Asia. The Carl Beck Papers in Russian and East European Studies No. 905. University of Pittsburg Center for Russian and East European Studies. 124p.

Mitrofanov, V.P., Dukravets, G.M. & Mamilova, R.Kh., 1988. Special features of the formation of the hydrocoenose in the Kapchagay reservoir on the River Ili. In: Impact of Agriculture on Biological Resources of Waterbodies in Kazakhstan, pp. 43–86. Izd. "Nauka" Kazakh SSR, Alma-Ata. (In Russian)

Mitrofanov, V.P., Dukravets, G.M. & Sidorova, A.F. 1992. Fishes of Kazakhstan. Vol. 5: Introductions and fisheries. Akad. Nauk Resp. Kazakhstan, Inst. Zoologii. 'Gylym', Alma-Ata. 462p. (In Russian)

Nikolaenko, V.A. 1988. Classification of water in reservoirs of Central Asia based on their chemical composition and their evaluation for irrigation use. Vodnye Resursy No. 2: 115–121. (In Russian)

Pavlovskaya, L.P. 1995. Fishery in the lower reaches of the Amu-Darya under the impact of water manipulation for irrigated agriculture. FAO Fisheries Circular. No. 894: 42–57. FAO, Rome.

Petr, T. 1992. Lake Balkhash, Kazakhstan. Internat. J. Salt Lake Research 1: 21–44.

Petr, T. 1995. Fisheries in irrigated areas of Central Asia. FAO Fisheries Circular. No. 894: 1–9. FAO, Rome.

Petr, T. 1998. Fisheries in reservoirs in the arid and semiarid zones of Asia. Internat. Rev. Hydrobiol. 83 (Special Issue): 591–598.

Petr, T. & Mitrofanov, V. P. 1998. The impact on fish stocks of river regulation in Central Asia and Kazakhstan. Lakes & Reservoirs: Research and Management 3: 143–164.

Pircher, W. 1993. 36 000 dams and still more needed. Water Power & Dam Construction, May 1993: 15–18.

Salikhov, T.V. & Kamilov, B.G. 1995. Ichthyofauna of the mid-Syr-Darya basin. J. Ichthyol. 35: 61–71.

Sanin, M.V. & Shaporenko, S.I. 1991. On the use of drainage waters for fisheries. Vodnye Resursy No. 4: 147–152. (In Russian)

Walker, K.F. & Yang, H.Z. 1999. Fish and fisheries in western China. FAO Fisheries Technical paper. No. 385: 237–278. FAO, Rome.

# THE USE OF IRRIGATION SYSTEMS FOR SUSTAINABLE FISH PRODUCTION IN PAKISTAN

**by**

**Nasim Akhtar**
**Aquaculture and Fisheries Research Institute, Pakistan Agricultural Research Council**
**Islamabad, Pakistan**

## ABSTRACT

The irrigation system of Pakistan is one of the largest integrated irrigation networks in the world, serving 14.4 million hectares of contiguous cultivated land. The waters of the Indus River and its tributaries feed the system. The salient features of the system are three major storage reservoirs, namely, Tarbela and Chashma on the River Indus and Mangla on the River Jhelum; 19 barrages; 12 inter-river link canals and 43 independent irrigation canal commands. The total length of the main canals is 58 500 km, with the field ditches running another 1.6 million km. Pakistan agriculture is predominantly irrigated. Hence the development of rivers for diverting water into canals is basically meant for irrigation supply. However, other uses such as effluent and waste disposal, watering livestock, fishing, recreation, transportation and washing and bathing are also on the increase. Crop production accounts for almost 70 percent of the agricultural gross domestic product (GDP). The major crops are wheat, rice, cotton, sugar cane and maize. Wheat is the leading food grain in Pakistan, and about 79 percent of the total wheat crop comes from irrigated fields.

Fish production in Pakistan during 2000 was 654 500 tonnes, of which 474 400 tonnes (about 72 percent) was caught from the sea whereas 180 000 tonnes was obtained from inland waters, 80 percent of which was captured from a network of rivers, irrigation canals and lakes. Only 20 percent of the inland fish production is contributed by aquaculture. The overall per capita fish consumption is low, about 2 kg/year. Fisheries are secondary users of reservoirs in multipurpose dams. Fisheries management in such reservoirs is still not very advanced. There is no effective data collection system which would assist with determining factors contributing to sustainable use of these reservoirs for fisheries. But it is realized that repeated releases of fish seed produced from hatcheries is an important measure for enhancing reservoir fish stocks. Canal systems have been given low attention in terms of realizing their potential for fish production. Active canals are rarely managed for capture fisheries and there is no cage or pen culture. However, abandoned irrigation canals are to some degree utilized for fish production. Flood control compartments are often stocked naturally with wild juvenile fish, including major carps and other food fish, during floods. Large areas of wetlands resulting from irrigation practices have considerable potential for fisheries development. To further improve the fisheries in waterbodies of irrigation systems will require better planning, improved data collection and evaluation, improvements in infrastructure, better coordination among fisheries organizations, and enforcement of fisheries regulations.

## 1. INTRODUCTION

The fish production propensity of a river system along with its allied structures such as reservoirs and irrigation canals is governed by a set of environmental parameters, especially the water and soil quality which in turn are the function of the geoclimatic conditions under which it exists. The following section provides information for a better understanding of the situation in Pakistan.

### 1.1 The geoclimatic features

Pakistan is a country of 135 million people. Situated in southwestern Asia it lies between latitudes 24º–37ºN and longitudes 61º–76ºE. The tropic of Cancer passes immediately south of the country. It covers an area of 796 000 km², bordering with Iran in the west, Afghanistan in the northwest, China in the north, and India in the east. Land and water features of Pakistan are given in Annex 1. The landscape includes high mountain ranges of the Himalayas, Karakoram and Hindu Kush with interspersed valleys, high snow-covered mountain peaks like K–2 (8 611 m) and Nanga Parbat (8 126 m), and seven large glaciers. There are vast irrigated plains in the Indus basin, stark deserts and rugged rocky plateaus.

Pakistan consists of two distinct physiographic provinces: the Western Highlands, and the Indus Plain resulting from the deposition of sediments by the Indus River and its tributaries. The Western Highlands extend from the Makran coast in the south to the Pamir Plateau in the north covering most of Balochistan, Northwest Frontier Province (NWFP), the Northern Areas and part of the Punjab Province.

The Indus Plain begins at the southern end of the Himalayas and the Salt Range and stretches to the Arabian Sea. The Indus Plain is the most important physiographic region since it comprises the main agricultural areas of the country. It covers an area of nearly 21 million hectares (mha). The aquifer underlying the Punjab delivers large quantities of water. The flat plain is largely made of alluvium, over 300 m deep, deposited by the Indus River and its tributaries, which flow through the provinces of Punjab and Sindh to the Arabian Sea near Karachi.

The Indus delta and its mangroves are the most significant features of the coastal area. The mangroves area is estimated to be the fifth largest in the world and serves as a nursery for a large variety of fish, shrimps and other marine life.

Pakistan has a large variety of climatic types ranging from temperate climate in the north to hot arid tropical in the south. The northeastern mountainous and submountainous areas receive more than 1 700 mm annual rainfall with the major share (1 000 mm) coming from summer monsoon. On the other hand, the extremely arid plains of southwestern Balochistan receive only 30 mm on the average during the whole year. Thermal regimes exhibit extreme diurnal, seasonal and annual variations. Temperatures fall as low as –26 °C over the northern mountains and reach up to +52 °C over the central arid plains. In the plains, temperatures of 42 °C are recorded at various stations in May-June. The climate of Pakistan is predominantly arid to semi-arid, with a small sub-humid to humid area in the north.

## 2. RIVERS, RIVER SYSTEMS AND AGRICULTURE CONSIDERATIONS

The Indus river system is the prime source of water in Pakistan and hence called the lifeline of Pakistan. The water resources of the Kharan desert (western Balochistan) and the arid Makran coast (southern Balochistan) are limited and the desert areas in the south (Thar and Cholistan) have no water resources.

The Indus River originates in southwestern Tibet from Lake Nagangla Ringco (31°40'N, 83°E) and makes a 3 280 km trip to the Arabian Sea, of which 2 480 km is in Pakistan. All the Pakistan major rivers, namely Kabul, Jhelum, Chenab, Ravi and Sutlej, flow into it. The system is like a funnel with a number of water sources at the top converging into a single stream, which flows first northeast, then turns southwest to finally drain into the Arabian Sea. The total length of the Indus River and its tributaries is estimated to be 63 000 km. Aided by a number of smaller tributary rivers and streams, these rivers supply water to the entire Indus Basin Irrigation System. The rivers have their origin at higher altitudes and derive their water mainly from snowmelt and monsoon rains. The catchment area of the Indus covers one million km² and contains several of the world's highest mountains.

## 2.1 Irrigation systems

The irrigation system of Pakistan started developing 150 years ago and since the Indus Water Treaty between India and Pakistan (1960) it is the largest integrated irrigation network in the world, serving 13.8 mha of contiguous cultivated land. The waters of the Indus River and its tributaries feed the system. The salient features of the system are three major storage reservoirs, namely, Tarbela and Chashma on the Indus, and Mangla on the Jhelum; 19 barrages; 12 inter-river link canals and 43 independent irrigation canal commands. The total length of main canals is 58 500 km with the field ditches running another 1.6 million km.

Diversion of river waters into canals is made through barrages, which are gated diversion weirs. The main canals in turn deliver water to branch canals, distributaries and minors. The water courses get their share of water through outlets in the irrigation channels. Distribution of water from a water course is effected through a time-schedule or "warabandi" under which each farm gets water for a specified period once a week. The time-share or "wari" is proportional to the farm area owned by a farmer under the command of the water course.

The system draws an average of 127.2 billion m³ of surface water each year for irrigation. Supplemented by an annual groundwater pumpage of some 51.6 billion m³, the average volume of water available at the farmgate is 3 685 m³ per ha. Approximately 3 million individual farms with an average size of about 5 ha benefit from this system.

### 2.1.1 Barrages

Five barrages (Chashma, Rasul, Qadirabad, Marala and Sidhnai) and a siphon were constructed. The siphon was constructed at Mailsi. There was also improvement and utilization of the existing barrages, Trimmu and Taunsa, and of Balloki Head Works.

### 2.1.2 Canals

The area served by each branch canal is called canal command and in the Punjab there are 24 canal commands, 5 in NWFP and 15 in Sindh and Balochistan. The total length of canals, their capacity, gross command area and the number of outlets in each province is given in Table 1.

**Table 1**
Irrigation canal system in Pakistan

| Province | Canal capacity ($m^3$) | Length of canals (km) | Gross command area (ha) | No. of outlets or watercourses |
|---|---|---|---|---|
| NWFP | 1 531 | 2 510 | 0.39 | 3 357 |
| Punjab | 3 371 | 36 862 | 9.70 | 49 332 |
| Sindh | 3 366 | 20 722 | 5.87 | 35 018 |
| Balochistan | 8.5 | 2 653 | 0.34 | 1 393 |
| Total for Pakistan | 7 326 | 62 648 | 16.51 | 89 100 |

The rate of growth in the farmgate water supply has declined in the past years. This will have serious implications for the country's future food security and while Pakistan possesses an intricate system of irrigation, its productivity and efficiency per unit water availability is perhaps the lowest in the world. This is due to reduced growth and water availability, as there has been no significant water storage development since the completion of Tarbela dam in the 1970s. The water losses are immense as 25 percent is lost from the canal head to the outlet and another 15 percent from the outlet to the farmgate due to poorly linked canals and water courses and inadequate operation and maintenance expenditure in the canal system; a rigid warabandi system and unplanned development of private tube wells is resulting in undermining of the groundwater resources.

The following link canals existed in the Punjab before the water treaty with India: Tripple canals, Upper Jhelum, Upper Chenab and Lower Bari Doab; Haveli canal taken out at Trimmu; Montgomery (Sahiwal) Pakpattan Link Canal; Balloki Sulemanki Link; Marala Ravi Link (MR); Bambanwala-Ravi-Bedian-Diplpur Link (BRBD); a sub-link connecting MR and BRBD.

Eight link canals were completed after the Water Treaty with India, including the Indus Basin Works under Rasul Qadirabad Link, Qadirabad Balloki Link, Balloki-Sulemanki Link No. 2, Taunsa-Panjnad Link and Chashma-Jhelum Link.

A description of link canals and their construction in chronological order are given in Annex 2.

### 2.1.3 Dams for power generation and water storage

The salient features of large and medium reservoirs of Pakistan are given below. (See also Table 2).

Warsak Dam
In 1960 a dam at Warsak on the Kabul River was constructed for power generation.

Mangla Dam
Under the Indus Basin Treaty, an earthen dam at Mangla on the Jhelum River, 100 m high at an elevation of 1 234 m and having length of crest equal to 3 352 m was constructed to store

6.6 billion m³ of gross storage with a live storage of 5.7 billion m³. The dam has five diversion tunnels 610 m long and 9.1 m in diameter to produce power. Its storage has submerged 27 125 ha. The main storage is in Khad and Poonch rivers and extends to Kanshi.

Tarbela Dam

An earth and rockfill dam, 147.8 m high with crest elevation at 1 565 m and length of 2 953 m was completed in 1975. Its gross storage is 13.2 billion m³ and live storage equals 11.1 billion m³. It has four tunnels each 14.8 m in diameter and 787 m long. A fifth outlet for irrigation has also been constructed. The main storage extends along the old river channel and a portion of the reservoir extends to Siran and to its tributary. It is estimated that the useful life of the storage is 50 years by which time 1.2 billion cubic meter storage will persist for ever.

**Table 2**

The salient features of large and medium reservoirs of Pakistan

| Dam | River | Year of completion | Area (ha) | Depth (m) | Drawdown (m) | Purpose |
|---|---|---|---|---|---|---|
| Mangla | Jhelum | 1967 | 26 667 | 91 | 51 | Irrigation Hydel |
| Tarbela | Indus | 1974 | 27 083 | 137 | 76 | Irrigation Hydel |
| Chashma | Indus | 1971 | 40 000 | 8 | – | Irrigation Hydel |
| Hub | Hub | 1981 | 10 417 | 46 | 19 | Irrigation |
| Khanpur | Haro | 1983 | 833 | 51 | 24 | Irrigation |
| Warsak | Kabul | 1959 | – | 48 | 8 | Irrigation |

(note: Hydel = hydroelectric)

### 2.1.4 Medium and small dams

Warsak, Tarbela, Khanpur, Baran and Tando dams are located in the NWFP. A large reservoir was created above Chashma Barrage. Mangla, Rawal, Simly and Namel reservoirs and 13 small dam reservoirs lie in Punjab. Several smaller reservoirs including Hub, Wali Tangi, Nari Bolan, Khushdil Khan and Spin Karez, are located in Balochistan. Gomel Zam and Kalabagh dams are still in the planning stage.

The Water and Power Development Authority (WAPDA), which is in charge of all larger dams in Pakistan, has identified fifteen more dam sites at Sehwan, Mancher Lake, in Chottari Depressions and Hamel Lakes. A few off-storage dams have been proposed, including Gariala, Dhok Pattan and Thal. (For the WAPDA fisheries activities in reservoirs see George, 1992, 1995).

### 2.1.5 Mini-dams

On the Potowar Plateau damming of watercourses to harness rainwater has created 500 impoundments. These perennial waterbodies are called mini-dams. Besides their irrigation use these multipurpose reservoirs are suitable for fish farming. Their size is generally less than 10 ha, hence they are easily manageable. Some of them already have aquaculture.

## 2.2 Purpose of the irrigation system

Pakistan's agriculture is largely based on irrigation. While the development of rivers for diverting water into canals is basically meant for irrigation supply, other uses such as effluent and waste disposal, livestock watering, fishing, recreation, transportation, washing and bathing are common.

### 2.2.1 The crop sector

Farm production is dominated by crop production, which accounts for almost 70 percent of agricultural GDP. The major crops are wheat, rice, cotton, sugar cane, and maize. Wheat is the leading food grain in Pakistan, and about 79 percent of the total wheat crop comes from irrigated fields. Rice is the second most important crop, and the two major rice-growing provinces are Punjab and Sind, which account for 54 and 36 percent of the area planted. Basmati rice, a high-value export crop and a foreign exchange earner, is grown chiefly in Punjab. Cotton, the most important non-food crop and largest export earner, occupies the second largest area of all crops grown. Besides production of lint, cotton seed accounts for 80 percent of the national oil seed production. Seventy percent of cotton is grown in Punjab. Sugar cane, yet another important cash crop, is grown over 70 000 ha. Maize is grown in both irrigated and rain fed areas and NWFP leads in maize production. The oil seed production has remained nearly stagnant for many years, despite the fact that the demand for vegetable oil has tremendously increased in recent years. The relative importance of coarse grains in the Pakistani diet has declined over time, as tastes have changed and per capita incomes have increased. However, grains such as sorghum, millet and barley constitute a major source of feed and fodder and their importance will increase further if the poultry sector continues to expand at its present rate. Gram accounts for about two thirds of the total acreage under pulses. Fruits and vegetables represent a dynamic segment of Pakistan's agriculture.

### 2.2.2 Seepages

The losses from water courses are between 10 to 20 percent with a maximum of 40 percent, half of which is due to evaporation losses while the rest seeps downward. The seepage water recharges the groundwater of the Indus plains. The seepage water has created problems such as waterlogging. To cope with this problem open surface drains and pumping of groundwater has been carried out on a massive scale.

### 2.2.3 Flow rates and seasonality of the use of irrigation and drainage canals

The base period of irrigation which represents the number of days for a crop season is 183 days for Rabi (winter crop season) and 182 days for Kharif (summer crop season). The total availability of water for irrigation is given in Annex 1.

### 2.2.4 Pumping system and fish-protection devices

Only two barrages have been provided with fish passes, however, they do not function properly. *Tenualosa ilisha*, a very popular anadromous species which migrates about 200 km from the coast upstream the Indus River for breeding, is blocked in its spawning migration by an ineffective ladder in the GM barrage. This has resulted in a drastic decrease in its stocks. The Chashma barrage fishway is also not efficient for most of the Indus River fish species. Mangla dam, which has no fish pass, blocks the *Tor putitora* upstream migration. Subsequently, this popular game fish is becoming rare.

No fish protection devices or refugia for fish are present in canals. Most pumping stations have no fish barriers at the intakes, and where they do exist, they do not work properly.

### 2.2.5 Water quality in the Indus, Jhelum and Chenab

The concentrations of total dissolved solids in these three major rivers of Pakistan are fairly low, averaging between 120 and 200 mg/litre. The concentration is lowest during the monsoon and increasing slightly during winter months. The chemical composition for the three rivers is given in Annex 3.

The Ravi and Chenab rivers' water vary in their dissolved salt concentrations, sediment load, as well as in concentrations of Cu, Fe, Mn, Zn, available phosphorus and organic matter content in the sediments. The water quality with respect to salinity is within the safest limits for irrigation. $Ca^{++}$ and $HCO_3^-$ are the dominant ions. The sediment load is, however, rising which indicates increasing erosion in catchments.

### 2.2.6 Use of agrochemicals/integrated pest management

Plant protection measures in Pakistan are largely limited to the use of pesticides. Undoubtedly, plant protection is an expensive input, but returns are also high. The use of integrated pest management is being slowly adopted.

# 3. FISH AND FISHERIES CONSIDERATIONS

## 3.1 Fisheries profile

### 3.1.1 Significance of the fisheries sector and its economic role

Pakistan is a predominantly agricultural country where about two thirds of its 135 million population is dependent, either directly or indirectly, on agriculture. Fisheries, a sub-sector of agriculture, plays a significant role in Pakistan's national economy and contributes towards the food security of the country. Contributing about 1.0 percent to the total GDP, which is equivalent to about 4.0 percent of the GDP of the agriculture sector, it absorbs 1.0 percent of the country's labor force. There are about 400 000 fishers, mainly fishing the sea. With their families they depend on fishing for their livelihood. The export of fish and fishery products during 1999–2000 reached 92 000 tonnes, this corresponding to a value of $US 139 million.

Recorded fish production in Pakistan during 2000 was 654 500 tonnes, of which 474 400 tonnes (about 72 percent) were from marine waters and 180 000 tonnes were from inland waters. Eighty percent of the freshwater catch came from wild resources of rivers, irrigation canals and natural lakes. Twenty percent of the inland fish production is contributed by aquaculture. Aquaculture in Pakistan benefits from irrigation, as about 75 percent of the fish ponds in the country are supplied with water by irrigation canals.

Collecting fishery freshwater statistics is highly inefficient so that official statistics fail to include a substantial amount of catches landed at a large number of unknown sites, including subsistence level catches by artisanal fishers and fish captured illegally. If all were to be included, the total catch would be much higher, perhaps twice as high as reported.

### 3.1.2 Major fish stocks, biodiversity and population decline/extinction

There are approximately 200 species of fish in freshwaters of Pakistan, of which a number are endemic. Substantial quantities of commercially important fish are caught from rivers. The inland commercial fishery encompasses about 30 species of which the significant ones are: *Labeo rohita, L. dyocheilus, Catla catla, Cirrhinus mrigala, C. reba, Channa striatus, C. marulius, Aorichthys (Mystus) seenghala, A. (M.) aor, Wallago attu, Rita rita, Bagarius bagarius, Tenualosa ilisha, Notopterus notopterus, Tor putitora, Schizothorax* spp., *Clupisoma nazirri* (in cold water streams only). The exotic species, which have managed to escape into the natural systems and have thrived there, also make a small contribution to the total landings from rivers. These include three species of tilapia, three Chinese carps, common carp and two species of trouts. (For lists of species for northern Pakistan, as well as for literature references, see Petr, 1999).

While none of the species are yet recognized as endangered at the national level, at least two of commercial importance, i.e. *T. putitora* and *T. ilisha*, may attain the status of endangered species soon if steps are not taken to conserve them. *Tor* migrates from floodplains to the Himalayan foothills for breeding and the creation of Mangla and Tarbela dams blocked its migration. *T. ilisha* needs to migrate upstream from the Arabian Sea to breed but ineffective fish ladders in the Guddu barrage on the Indus River hinder the migration. Many northern cold water hill stream loaches of the genus *Nemacheilus* and *Triplophysa* are threatened; at least 10 species of cyprinids are rare and vulnerable; cold water indigenous snow trout has a limited range of distribution and is being outcompeted by the introduced exotic trout and thus is of special concern. A number of studies have been done in Pakistan to identify the fish and their natural distribution, however, a large number of species have not been described in terms of their habitat requirement and population structure. The issue becomes more important when we look into the matter from the perspective of the range of climatic and habitat diversity prevalent in the country. Moreover, human activities in the last five decades, such as logging, agriculture, highway construction, diversion and regulation of rivers through creation of dams, barrages and other waterworks, have greatly decreased the number of suitable habitats. Habitat loss has also resulted from organic and other types of pollution, including discharges of toxic substances. Consequently, those species with limited distribution and critical habitat requirements may disappear from inland waters of Pakistan due to habitat loss or modification. In-depth research is imperative, not only to determine distribution, taxonomic status, population size and habitat requirements of these species but also to devise a rational conservation programme for their protection. The research should culminate in a programme of establishment of sanctuaries, nurseries for salvage of juveniles during floods, and hatchery techniques for artificial breeding of vulnerable species.

### 3.1.3 Present use of irrigation systems for fish production

The inland fisheries of Pakistan consists of three sub-sectors: traditional capture fisheries in rivers, lakes, reservoirs, barrages and other waterbodies; sport or recreational fisheries, primarily in the NWFP and the Northern Areas; and aquaculture in small and large ponds.

The current status of capture fisheries is primarily the result of the prevalent institutional framework for inland fisheries management in the country.

The institutional arrangements for fisheries in Pakistan are quite extensive. Inland fisheries is primarily a provincial responsibility. The provinces have jurisdiction over all the waterbodies within their province with the exception of large reservoirs which are managed by WAPDA. Capture fisheries is subject to the Fisheries Ordinance/Acts of the provincial governments with respect to conservation.

Traditional inland fisheries is managed by an auction system and limited licensing of natural waterbodies, for which purpose the rivers, canals, barrage reservoirs and head works are compartmentalized arbitrarily and the fishing rights for each compartment is annually auctioned publicly by the respective Fisheries Department after wide publicity. Likewise, fishing rights for the large and medium-sized man-made reservoirs on rivers are auctioned by WAPDA. Lakes, abandoned canals and dhans (usually large waterbodies created besides canals due to earth excavation and filled by seepage water from canals) are sold by the custodian Fisheries Department to the private sector after competitive marketing. The limited licensing, however, is subject to various regulatory rules, such as closed seasons, which basically are aimed at sustaining the fish productivity and conservation of the species. Whereas the auction system is quite well developed, the departmental capabilities and infrastructure for implementation of such rules are quite inadequate. The management of the waterbodies under this system is not efficient as the magnitude of the resource is not well known. There is a dearth of information on the maximum sustainable yield (MSY) and on fish production. The indicators, however, suggest that the fishing pressure is reaching the ceiling of the MSY. Management needs to pay more attention in order to insure that the resources are not overexploited. The sport and recreational fisheries are managed by licenses, particularly the coldwater fisheries in the riverine systems of the Northern Areas. Fish stock protection and enforcement are major management problems in this sub-sector.

Another reason for neglect of capture fisheries is that most of the expenditure at state level has shifted towards aquaculture rather than capture fisheries in the last decades. The rationale is that aquaculture provides a way of enhancing natural stock and therefore benefits the capture fisheries. However, the freshwater aquaculture is still relatively small scale, with a total farming area of about 50 000 ha. But as there is an element of ownership involved in aquaculture as compared to no ownership in capture fisheries, there is the likelihood that aquaculture will grow fast. This is much more so because of high profitability in aquaculture, feasible environment, and top government support for its growth.

#### 3.1.3.1 Fisheries in large reservoirs

The salient features of reservoir fisheries in Pakistan are:

- Fisheries are secondary users of reservoirs hence in multipurpose dams fisheries have a lower priority.
- The little research already done on reservoir limnology and fisheries is still not adequate for developing thorough management strategies. WAPDA has improved fisheries management in some of the major reservoirs.
- At the departmental level there is no effective data collection system with respect to factors contributing to sustainable fisheries in reservoirs. Training of manpower in the analysis and management of data is particularly lacking.

- The impact of stocking on yield enhancement is not fully understood. In this regard so far no attempt has been made to develop an optimal stocking model for each reservoir. Predictive models for estimating potential yields need an adequate database, which is lacking.
- Human impacts in watersheds, such as deforestation and discharge of pollutants into streams are but a few of the hazards which have not so far been properly addressed.

#### 3.1.3.2 Fisheries in canals

Although dams, lakes and rivers as components of irrigation schemes have been a big source of fish, the canal systems have been given less attention in terms of realizing their potential for fish production. The active canals, i.e. those with water, are rarely managed for capture fisheries and there is no cage or pen culture. A number of inherent management problems exist in irrigation systems, which results in underutilization of these resources for fisheries. Saline water associated with an irrigation system, as present in drainage canals, provides a suitable environment for fish like tilapia and some crustaceans but no effort has been made towards making use of this resource. Aquatic weeds in irrigation canals could either be harvested and used as fish food or they could be grazed by grass carp but both approaches are yet to be implemented in Pakistan. No attempt has been made to follow the regime of water flow in irrigation canals, the knowledge of which is essential if the canals are to be considered for aquaculture. Pakistan is, however, doing well in utilization of abandoned irrigation canals for fish production. After slight modification these waterbodies are treated as fish ponds. Most of these impoundments have extensive culture systems in which stocking is rarely followed by other management inputs.

#### 3.1.3.3 Flood control compartments and fisheries

Along the Indus River, for about 150 km between the Guddu and Sukkur Barrages, there are numerous flood control compartments with water control gates. They range in size from 10 to 5 000 ha each. Under normal operating procedures, these compartments are filled during the rainy season and subsequently drained a few months later after the floods have subsided. However, the peak flood period also corresponds with the peak of fish spawning. The flood control compartments are often stocked naturally with wild juvenile fish, including the Indo-Gangetic major carps and other food fish. Unfortunately, at the time the compartments are drained a few months later, the fish are still of a fairly small size. However, if these compartments were to be managed as a fish pond and the fish allowed to grow larger, there would be a good potential for better use of an infrastructure that already exists. There are over fifty of these flood control compartments and if only a fraction of them were managed to produce one to two tonnes of fish each year, the potential for increased fish production would be substantial, especially considering the relatively minor investment required.

#### 3.1.3.4 Mini-dam reservoir fisheries

In barani areas, a large number of mini-dams have been created to harness surface runoff, particularly after rains. These are generally created on non-perennial streams and are used for small irrigation schemes. These are unique new waterbodies with a mix of the characteristics of a pond and a reservoir. Most of these mini-dams are privately owned, constructed with government subsidies, and very few of them are used for fish production mainly due to non availability of appropriate fish production methodology for these mini-dams.

#### 3.1.3.5 On-farm pond aquaculture

In a given agriculture system the main aquaculture setups are in the agriculture farmer's residual farmland. There is also the tendency to build fish farms on marginally productive agricultural land of the farmer. These farms are supplied with either tubewell water or from the irrigation canal. As the dependence on the canal water in fish farming systems is very high, one can consider these ponds to be an integral part of the irrigation system. The management requirements of these canal water-dependent ponds are different due to the high levels of silt in the water and the presence of undesirable fish species entering the pond with the canal water despite all precautions taken to prevent this.

#### 3.1.3.6 Fisheries in residual water storages

Terminal lakes and residual waterbodies have not yet been surveyed for their fish production potential and no development and management programme has been formulated for their use.

#### 3.1.3.7 Wetlands: fisheries perspective

Pakistan is endowed with immense areas of wetlands representing unique and diverse ecosystems. So far, there is no special management technique or plan for utilizing these bodies as a fisheries resource, although recent surveys have concentrated on assessing their potential. Three wetlands have been declared Ramsar sites for preservation and conservation purposes. A list of the important wetlands of Pakistan is presented as Annex 4. For fisheries purposes some of these wetlands are listed under the synonym of natural lakes.

Natural lakes cover an area of 109 780 ha. Some are high altitude coldwater lakes and are therefore well-suited for trout. Most warmwater lakes are in Sind Province and are a great source of fish. A canal feeds Lake Mancher in Dadu District from Lake Kalri. This lake was considered to be one of the best lakes of Sind, but its productivity has diminished due to overfishing and excessive aquatic weed growth.

Lake Kalri is situated about 100 km from Karachi and has two parts: one known as Lake Sonahri and the other as Lake Kalri itself. It has an excessive growth of weeds. Lake Haleji is situated 88 km north of Karachi and is fed by a canal from Lake Kalri. Fish landings from these lakes have been declining.

Many small lakes exist in Thatta and Sanghar Districts and they support good capture fisheries. Lake fisheries in Sind are regulated by the Department of Fisheries (Sind) and their fisheries management plan is mainly regulatory in nature.

#### 3.1.3.8 Fisheries and aquaculture in waterlogged areas

The extent of waterlogged areas is given in Annex 1. Of the millions of hectares of waterlogged areas created due to massive irrigation systems, practically no attempt has been made to use these waterbodies for fish culture. An interesting phenomenon observed during the current prolonged drought in Pakistan is that the water table in many waterlogged areas has gone way down due to acute shortage of water in the canal systems.

### 3.1.4 Fish marketing

Both marine and inland fish are marketed in Pakistan. The per capita fish consumption is a low 2 kg/year but there is an increasing trend. The per capita consumption, however, varies from place to place. For instance, in coastal areas it is 12 kg/year. The coastal provinces are supplied mainly by marine fisheries while the inland provinces are mainly supplied by inland fisheries. Marine fishes are distributed by train from Karachi (Sind) to major markets in other cities such as Lahore, Multan, Rawalpindi (all in Punjab) and Peshawar (NWFP). Inland fishery products are marketed in all major markets of the urban centers. The fishers bring their catch by boats to collection points on the shore of the lake/reservoir/barrage or river. There the contractors/collectors collect the fish, place them in woven baskets or burlap bags with crushed ice and then transport these, on a daily basis, by public or private transport to the major markets called "fish arrhat". The fish is auctioned here, based on grading by weight/size, species and quality. The availability of fish in the retail market is not only limited but is characterized by irregular supplies, low number of vending slots and inferior facilities for storage, distribution and marketing.

Fish weighing 2–3 kg fetch prime prices in the market. The price is also dependent on the degree of freshness and the market demand/supply situation. The market itself is seasonal with best prices during the coolest winter months. Other factors which affect fish price include location of the market (city, for instance), species of the fish, closure of canals for removal of sand (for a month during the first calendar year).

It is estimated that post-harvest losses from river capture fisheries are 20 percent. Inferior handling and distribution, inadequate infrastructure and poor marketing practices characterize the marketing system.

## 3.2 Management measures for increasing fish production in various types of inland waterbodies

Although the capture fisheries management in rivers, reservoirs and canals has been improving, there is still much to be done. There is a great scope for improvement in planning, data collection, infrastructure, and for better coordination among different fisheries organizations involved with the same resource at the macro- as well as micro-level. The large dams are primarily built for hydroelectric power generation and irrigation. To operate these systems specialized departments are involved. However, fisheries are still given very little power to interfere with the water distribution pattern which is dictated by the major uses. In WAPDA, its own Fisheries Department is responsible for fisheries management of the large water storages serving irrigation and hydroelectricity production. For at least 20 years this department has been implementing a programme of fisheries development in these reservoirs. The fisheries beyond these reservoirs, both up and downstream, remain beyond the WAPDA jurisdiction. Rivers are under the provincial fisheries departments, and the barrages and irrigation canals are under the provincial irrigation departments, with fisheries in them being the responsibility of provincial fisheries departments. There is no body at the state level to guide activities of these fisheries organizations to devise a rational fisheries management programme for waterbodies serving irrigation purposes.

Several of the known management inputs are being used in Pakistan with the objective of sustaining the current fish production levels from the river/irrigation systems. Practices such as species introductions, artificial recruitment, introduction of new species and regulatory measures including closed seasons are not properly understood and hence these are applied as and when deemed suitable but without working out their rationality and strategic planning. It can be said that the overall advancement and development towards understanding and follow up action plans for irrigation system fishery-cum-aquaculture is still very much in its infancy, in Pakistan as well as in some other countries in the arid zone of Asia.

### 3.2.1 Fish introductions and stocking

The previous two decades have seen the introduction of the herbivorous cyprinids silver carp and grass carp, of common carp and very recently of bighead carp and three cichlids, *Oreochromis niloticus*, *O. aureus* and *O. mossambicus*. These species were introduced into Pakistan for pond aquaculture to fill the vacant trophic niches of the major carp-based polyculture system in warm water regions. Insofar as introduction of grass and silver carp is concerned, their stocking in open water has posed not much of a problem as in their environments the species meet with only indifferent breeding success. The interrelationship between indigenous carps and Chinese carps in open waters of Pakistan is not yet fully understood, but it is known that in India silver carp is a severe competitor with catla. Stocking enhancement in open systems needs careful follow-up monitoring and evaluation of the introduced exotic species in terms of both their growth rate and reproduction.

The river system modified for hydroelectric power generation, and its allied canal system, are the dominant areas for further development of inland capture fisheries. Of the total inland fish production in the country about 25 percent comes from six reservoirs.

It is realized that enhancement of fish production through repeated releases of stocking material produced in hatcheries is an important management measure in reservoir fisheries. Such a stock enhancement technique supplements or sustains a population of species that already exists or that was introduced before. Some reservoirs in Pakistan have been stocked for a number of years. These stocking programmes have been implemented without properly understanding the causes behind some declines in fish production, and with only a limited knowledge of the inherent productivity of some reservoirs. This has resulted in failure of some stocking programmes in the past because factors such as overfishing, poor fisheries management and degradation of aquatic environment have not been simultaneously addressed.

### 3.2.2 Other measures applied to achieve sustainable fish production in a variety of irrigation waterbodies (e.g. fisheries laws and regulations and their enforcement)

Experience has shown that uncontrolled fishing and destructive fish capture methods deplete fishery resources, with logical consequences for those who depend on fishing. To control illegal fishing a number of laws and regulations already exist.

The Indian Fisheries Act, which came into being in 1897, provides a basis for the current Fisheries Act of Pakistan; it has been adopted with some modifications by each of the provinces of Pakistan. Provisions against public nuisance such as pollution exist in the Pakistan Penal Code, Criminal Code and State and Local Acts, by which legal powers are conferred upon the State. Inspectors of Factories control proper treatment of industrial effluents before disposal. Such provisions have not always prevented the pollution.

Restrictions have been imposed on the use of certain nets during specified parts of the year as well as on the mesh size. Closed season, i.e. restriction on fishing during a certain period of the year, mostly covers the period from 15 April to 15 August. In the 1960s, the Government prohibited catching of rohu, mrigala, mahseer and catla smaller than 25.4 cm. Certain areas have been declared as "protected waters" or sanctuaries and closed to fishing. Restrictions on the sale of illegal–size fish and issuing of fishing licenses are other regulatory measures.

While the Fisheries Act and the legislation exist, the machinery for the enforcement of the regulations in most cases is so inadequate that the objectives are hardly fulfilled. Despite legal prohibition, the capture and destruction of brood fish and juveniles in large quantities are commonly practised all over the country and are largely responsible for the impoverishment of fish stocks. Legislation based on empirical knowledge is of doubtful utility. There is thus a need to discard legislation of suspect importance and instead formulate and enforce regulations based on the analytical approach of unit populations of different species of fish in rivers and allied irrigation structures such as reservoirs, barrages, etc.

### 3.3 Use of agrochemicals and biodegradability

Organochlorines, organophosphates, carbamates and synthetic pyrethroids are the common agricultural insecticides. All insecticides vary in degree of biodegradability and toxicity to fish. Fenvalerate, one of the pyrethroids effectively used for pest control in cotton and a variety of other crops in Pakistan, is highly toxic to fish. Bioaccumulation of xenobiotics in aquatic species is increasing alarmingly and this is also increasing the potential human health hazard. A variety of fish species show absorbance and accumulation of many contaminants such as pesticides and bactericides, polychlorinated biphenyls and heavy metals. Due to the lipophylic characteristics of many of the insecticides, they (especially pyrethroids) are absorbed by fish, even at low concentrations in the water. The toxicity caused to fish may not be immediately harmful, but most of the insecticides disturb haematological and metabolic processes by inhibition/activation of enzymatic activities and cause tissue damage.

### 3.4 Other pollutants

Degradation of rivers, reservoirs and canal systems has been on the increase due to the rapid pace of industrialization, poor environment management in the catchments and a variety of other factors. Apart from the direct entry of industrial, municipal and thermal wastes, the pollution load carried by the upstream rivers also accumulates in the reservoirs. The environmental degradation in reservoirs and barrages is caused mainly by the waste discharge from industrial, municipal and agricultural sources. High rate of siltation due to poor catchment management also affects the biological productivity. The need to reduce the amount of pollutants reaching rivers, lakes and reservoirs in order to mitigate their impact on fish and fisheries, is hardly receiving any attention in Pakistan.

## 4. INTERSECTORAL COOPERATION TO OPTIMIZE THE USE OF IRRIGATION SYSTEMS FOR FISH PRODUCTION

Water engineering has played an important role in Pakistan agriculture for over 5 000 years. The agriculture irrigation schemes have impacted the natural river systems by transforming fluvial conditions to a combination of fluvial and lacustrine conditions, with additional unique features of their own in adjoining canals in terms of spatial variations of certain habitat variables. The ecology of reservoirs is radically different from that of the parent rivers. Dams alter river hydrology both up- and downstream. Another unique feature of reservoirs and irrigation canals, which makes them distinctly different from their natural counterparts, is the water renewal pattern marked by swift changes in levels, inflow and outflow. The complexity of managing an altered water regime for sustainable fish production, therefore, is more complicated than managing a natural system. It is no longer a capture fishery alone; to achieve a sustainable and reasonable fish production which can support fisheries requires a mix of capture and culture fisheries. In turn, this requires management inputs which consider a multitude of factors, which can only be discussed with the full participation of all users of the water resources, such as irrigation and hydropower engineers, crop production specialists, and other parties involved in the use of water. Socio-economic aspects must be included in any fishery management considerations pertaining to the full development of fishery potential in waterbodies of irrigation systems.

## 4.1 Suggestions for improvement

Keeping in mind the complexity of the social, economic and ecological factors involved in the formulation of a sound management policy and in the implementation of a programme of sustainable fish production development from irrigation systems, the following steps need to be considered.

### Availability of data

The most important thing is the availability of data on the resource and the rate of fish stock exploitation.

The inadequacy of knowledge about the resource biology, different habitats, stocks, behavior and dynamics of individual fish species, population structure and recruitment pattern in relation to river modifications/development in the country needs to be addressed. A sound data base is essential for comprehensive analysis of biological and environmental factors required for the formulation of an integrated fisheries management plan for various types of waterbodies representing an irrigation system.

All large, medium, small and mini-dam reservoirs need to be classified according to their morphometric, physical and biological parameters on the basis of time-series data. This data should be analysed under a unified theme and a goal to which the data is harnessed under a holistic approach. Countries sharing the resource should join in this process. The process will facilitate rationalizing efforts for sustainable use for fisheries development of reservoirs and other irrigation system waterbodies.

### Better use of irrigation and drainage canals for fish production

Since fish production from irrigation systems is generally extractive in nature, selective stocking either for correcting imbalances in species spectrum or to fill vacant ecological niches assists in enhancing fish production. Wherever possible one or more of the following enhancement measures should be applied, together with conventional fisheries management practices. The implementation strategy, however, anticipates a profound knowledge of the system, its dynamics, and follow-up monitoring and evaluation of the stock enhancement programme in the waterbody where applied.

The following are some enhancement strategies:

- Stocking in suitable portions of irrigation systems to improve recruitment, altering fish assemblage structure to favoured species or to maintain such species that would not breed naturally in the system.
- Introduction of new/exotic fish species, in compliance with the Code of Practice, to exploit underutilized habitats
- Engineering the environment through provision of structures such as fish ladders to enable migration to improve levels of reproduction, etc.
- Systematic removal from the system of unwanted fish species which either compete with or predate upon target species.
- System modification for improved and reliable availability of nutrient-rich water from irrigation channels to feed fish ponds.

**Cage and pen culture**

Introduction of cage culture should be considered for suitable waters of irrigation systems. Pen culture development is encouraged where the morphology of rivers and reservoirs allows such structures to be installed.

**Control of water-borne diseases using fish**

Many rivers and irrigation systems have transboundary connections which facilitate the dispersion of fish diseases. What is needed is fish disease legislation, and capacity building through training and infrastructure development.

**Mitigating effects of pollution**

The existing legislation needs to be enforced, the environmental protection agency strengthened, and pollution "hot spots" ameliorated by implementing a rational management plan.

**Non-utilized waterbodies**

- Establish a scheme to study the different options for fish production in irrigation canals, especially through aquaculture.
- Use grass carp in canals partially choced with weeds for attaining dual benefits of commercial fish production and weed control.
- Large water areas created by waterlogging need to be assessed for their suitability for fish culture.
- Appropriate technologies for fish culture under fluctuating salinities need to be identified.
- Abandoned canals, flood compartments and seepage tanks besides canals are potential habitats that could be harnessed judiciously for aquaculture.
- Mini-dams offer suitable environment for fish production in a multiple-use scenario but this will require intensive experimentation for optimum resource utilization.
- Natural lakes such as Mancher in Pakistan, which are continuously fed by canal water, are unique systems offering challenges for fisheries through blending fish culture with capture fisheries.

- River floodplains, with their diversity of lakes and other types of backwaters, can be more efficiently used for fish production that they are at present.

**Fish production potential**

The following activities may be considered as steps to enhance fisheries in waterbodies of irrigation systems:

- Models tested in other parts of the world should be tested through a validation process to assess their suitability for improving fish yields.
- Factors limiting development of well-functioning fisheries cooperatives, including those slowing down the pace of improvement of the social welfare, particularly of small-scale operators, need to be reappraised. Community participation should be a viable option for improving the current fisheries management efforts.
- Constraints to fisheries technology, distribution and financing need in-depth evaluation.

## 5. RECOMMENDATIONS

Most of the following recommendations for improvement of fisheries in waterbodies serving irrigation focus primarily on the region, especially on the need for regional cooperation. Some recommendations are pertinent to problems in the individual countries.

- A comprehensive national survey for assessment and analysis of fishery resources in irrigation systems including rivers, reservoirs and canals is essential for developing sound management strategies.
- Development of a regional fisheries resource information system for creation of a regional database for waterbodies represented in irrigation systems, is recommended. This to be established in parallel with the strengthening or establishment of a similar information center in each participating country.
- Initiation of a collaborative international fish conservation programme including conservatory aquaculture to safeguard the most vulnerable species.
- Sharing of knowledge and information for increasing fish catches through improved harvesting technology particularly for reservoirs.
- Organization of an international workshop on fish production potential of irrigation and drainage canals.
- A collaborative regional research project for development of saline inland water aquaculture.
- Capacity building at national level for assessment of the impact of global change on inland fisheries and environment.
- Regional programme of research capacity building and linkages with international research organizations.
- A regional scientific study programme to address the problem of pollutants arising from irrigated agriculture, including their speciation, adsorption of pollutants on sediments, and impacts on fish, especially in drainage canals and waterbodies established from residual/drainage waters.
- Governments should seriously consider establishing protected zones in freshwaters for fish species endangered by modifications of natural waterbodies by irrigation structures.

## REFERENCES

George, W. 1992. Review of inland fisheries of Pakistan. FAO Fisheries Report No. 458, Suppl.: 47–81. Rome, FAO.

George, W. 1995. Review of fishery management in Water and Power Development Authority (WAPDA) reservoirs in Pakistan. FAO Fisheries Report. No. 512, Suppl.: 141–153. Rome, FAO.

Petr, T., 1999. Coldwater fish and fisheries in Pakistan. FAO Fisheries Technical Paper. No. 385: 122–137. Rome, FAO.

## ANNEX 1

Pakistan's resource base (source: WAPDA, Pakistan, and Pakistan Economic Survey)

135 million people (33 percent urban population)

### LAND

| | | |
|---|---|---|
| Geographical area | = 79.6 million ha (mha) | |
| Surveyed area | = 59.3 mha | |
| Area under cultivation | = 27.6% | (21.87 mha) |
| Crop area irrigated | = 22.6% | (17.99 mha) |
| Rainfed agriculture area | = 4.97% | (21.87 mha) |
| Forest | = 4.5% | (3.6 mha) |
| Culturable waste | = 11.7% | |
| Rangelands | = 59% | |

### WATER

| | |
|---|---|
| Indus River System with 5 tributaries | 110 000 km |
| Largest contiguous irrigation system in the world | 63 000 km |
| Generous water resources (especially during 70–100 summer days) | 168 billion $m^3$ |
| Water diverted from rivers to canals/year | 117.6 billion $m^3$ |
| Water reaching cultivable land | 44 billion $m^3$ |
| Water draining into the Arabian Sea | 46.9 billion $m^3$/year |
| Total water storage capacity of Pakistan | 20 billion $m^3$ |
| Potential for enhancement of storage capacity on the Indus River system | 70.75 billion $m^3$ |

### WATERLOGGED AREA

| | |
|---|---|
| Water table 15 cm | 2.2 mha |
| Water table 152 cm | 6.7 mha |

# ANNEX 2

Construction of canals (in chronological order)

**Earliest canals**

1351 Feroze Shah Tughlaq constructed Western Jamna canal.
1568 Shah Jahan improved Jamna canal in 1568 by the order of the Emperor Akbar.
1626 With the falling power of the Moguls Jamna canal continued to deteriorate.
1633 Shah Nehr was constructed to supply water to Shalimar Gardens from the Ravi River.
1817 The British started the improvement of western Jamna canal.
1846 After the annexation of the Punjab canals were improved.
1859 The Bari Doab canal constructed.
1872 Sirhind canal constructed.
1878 A severe famine occurred in Northern India so Sidhnai canal constructed in 1886.
1892 Lower Chenab canal built.
1901 Lower Jhelum canal constructed.
1855 Kabul canal improved.
1891 Lower Swat Canal opened.

**Canals constructed between 1900 and 1930**

- During this period triple canals such as Upper Jhelum, Upper Chenab and Lower Bari Doab. were constructed in the Punjab.
- The eleven Sutlej Valley canals were taken out of the four barrages constructed on the Sutlej canal system of the Punjab.
- Upper Swat Canal was taken out at Amandara in NWFP.
- In Sindh Sukkur Barrage was completed, including seven canals.

**Canals constructed between 1930 and 1960**

- Thal canal taken out of Indus at Kalabagh.
- Haveli canal opened in 1939, taken out at the confluence of Jhelum and Chenab.
- Taunsa Barrage in 1959 to provide weir controlled supplies to D.G. Khan and Muzaffargarh canals.
- Bambanwala-Ravi-Bedian-Dipalpur link by widening the existing Raiya Branch and providing a siphon through the Ravi to serve Bedian and Dipalpur.
- Balloki Sulemanki lined link.
- Kotri barrage was constructed to divert 1 171 $m^3$/sec into the four canals, three on the left bank and one on the right.

**Canals constructed during 1960 to 1970**

- Guddu barrage was completed in 1963 out of which two canals from the right and one from the left were taken out.
- In NWFP high-level canals were taken out of the Kabul River above Warsak dam. The left bank canal and the right bank canal. The right bank canal bifurcates into gravity canal and lift canal.
- On the Kurrum a weir was constructed and a canal was taken out. An earthen dam 38.7 m high, with a storage capacity of 106 800 000 $m^3$, was also constructed.

**Link canals existing before the Water Treaty with India**

- Triple canals, Upper Jhelum, Upper Chenab and Lower Bari Doab constructed between 1905 to 1915.
- Haveli canal taken out at Trimmu in 1937.
- A link called Montgomery (Sahiwal) Pakpattan to carry 24.7 $m^3$/sec was constructed.
- Balloki Sulemanki Link.
- Marala Ravi Link (MR).
- Bambanwala-Ravi-Bedian-Diplpur Link (BRBD).
- A sub link connecting MR and BRBD.

**Indus Basin Works**

Link canals

- They included eight links constructed for Indus Basin Works. These were: Rasul-Qadirabad Link, Qadirabad-Balloki Link, Balloki-Sulemanki Link No.2, Taunsa-Panjnad Link and Chashma-Jhelum Link.

Barrages

- Five barrages and a siphon constructed. These included Chashma, Rasul, Qadirabad, Marala and Sidhnai. A siphon was constructed at Mailsi.
- Improvement and utilization of the existing barrages. These included Trimmu and Taunsa Barrages and Balloki Headworks.

## ANNEX 3

Chemical analysis of water of the rivers Indus, Jhelum and Chenab (meq/litre)

| Sample (month) | Ca | Mg | Na and K | $HCO_3$ | Cl | $SO_4$ | Total cations and anions | $SO_2$ | TDS mg/ litre | EC µ mhos/ cm | pH |
|---|---|---|---|---|---|---|---|---|---|---|---|
| January | 1.62 | 1.18 | 0.34 | 1.7 | 1.54 | 0.9 | 3.14 | – | 192 | 296 | 7.7 |
| June | 1.50 | 0.59 | 0.42 | 2.04 | 0.19 | 0.25 | 2.51 | 6 | 160 | 245 | 7.5 |
| January | 1.39 | 1.04 | 0.23 | 1.7 | 0.09 | 0.87 | 2.66 | – | 160 | 250 | 7.7 |
| June | 1.83 | 0.18 | 0.66 | 2.0 | 2.0 | 0.19 | 2.67 | – | 160 | 250 | 7.4 |
| January | 1.62 | 0.62 | 0.81 | 2.02 | 0.19 | 0.84 | 3.05 | – | 180 | 290 | 7.3 |
| June | 1.10 | 0.61 | 0.40 | 1.10 | 0.2 | 0.80 | 2.11 | 4 | 124 | 195 | 7.4 |

Note: $CO_3$ and boron nil, $NO_3$ traces.
Source: Pakistan Council for Water Resources Management

## ANNEX 4

Wetlands of Pakistan

| Name | Area (ha) | Elevation (m a.s.l.) |
|---|---|---|
| Band Khushdil Khan | 1 296 | 1 460 |
| Hamun-i-Lora | 2 100 | 1 100 |
| Hamun-i-Mushkel | 1 800 | 1 000 |
| Marav Lake | 2 000 | 990 |
| Zangi Nawar | 2 070 | 975 |
| Indus Game Reserve | 81 000 | 700 |
| Indus Waterfowl Refuge | 3 774 | 500 |
| Head Islam | 3 132 | 139 |
| Kharrar Lake | 235 | 180 |
| Lal Suhanra Lake | 1 935 | 115 |
| Taunsa Barrage | 6 567 | 139 |
| Beroon Kirthar Canal | 1 000 | 70 |
| Charwo Lake | 100 | 40 |
| Drigh Lake | 182 | 50 |
| GhauspurJheel & Sindhi Dhoro Lake | 600 | 70 |
| Hamal Katchi Lake | 1 200 | 50 |
| Indus Dolphin Reserve | 44 200 | 50 |
| Khango Lake | 500 | 50 |
| Khinjar Lake | 13 468 | 70 |
| Khipro Lakes | 30 000 | 40 |
| Langh Lake | 19 | 50 |
| Mahboob Lake | 100 | 50 |
| Manchar Lake | 6 000 | 35 |
| Nara Canal | 30 000 | 70 |
| Phoosna Lakes | 160 | 40 |
| Pugri Lake | 300 | 50 |
| Sadhori Lake | 1 000 | 60 |
| Sanghriaro Lake | 380 | 60 |
| Soonhari Lake | 245 | 60 |
| Tando Bago Lake | 900 | 40 |

Source: WWF, Pakistan

This list, still incomplete, is based on GIS studies currently under way in Pakistan.

# THE USE OF IRRIGATION SYSTEMS FOR SUSTAINABLE FISH PRODUCTION IN TURKEY

by

**R. Celebi**
**Ministry of Agriculture and Rural Affairs, Tarim II Mud.**
**Kontrol Sibesi, Sanlirfa, Turkey**

## ABSTRACT

While the marine fish stocks in the seas surrounding Turkey are gradually diminishing, inland water resources offer a major scope for development of freshwater capture and culture fisheries. The potential has been enhanced by the rapid construction of many dams of irrigational and multipurpose character. The GAP project in south-eastern Turkey aims at eventually irrigating 75 000 $km^2$ of land. By 1998, irrigated areas in the GAP had already reached 299 554 ha. GAP is a multipurpose river basin project using the water of the Euphrates and Tigris and involves the construction of 22 dams The Euphrates reservoirs will have a total water surface area of 125 175 ha, and those on the Tigris 73 298 ha. In 1998, capture fisheries harvested 1 842 tonnes, and aquaculture produced 290 tonnes in the GAP region. Regarding fisheries, the existing fish stocks need better protection and rational management, and the local fishermen/farmers should be trained in capture fisheries methods and in aquaculture applicable in a variety of waterbodies of irrigation systems. The existing fishery regulations need to be adapted to the GAP region, which differs climatically from other regions of Turkey. This concerns especially closed seasons for individual fish species. Dam managers responsible for water releases should be made aware of the water requirement for fish spawning periods.

## 1. INTRODUCTION

Turkey is a large peninsula covering over 800 000 $m^2$. It is surrounded by four seas, the Mediterranean, the Aegean, the Black Sea, and the Sea of Marmara. From this favourable geographic position its people have had access to marine fishery resources for centuries. However, with the current population of 68 million, a sixfold increase in 75 years, and a similar growth in the neighbouring countries, the marine fish stocks have been put under increasing pressure. Turkey's marine capture fisheries have declined to about 500 000 tonnes per year.

Turkey, being the largest country in south-western Asia, has considerable agricultural potential. Twenty-eight million ha of agricultural lands, 21.5 million ha of pastures, 83 million heads of livestock all depend on the rainfall and irrigation. In terms of inland water potential, Turkey has 36 rivers with a total length of 178 000 km, over 200 lakes covering 9 000 $km^2$, 142 reservoirs with total water surface area of 2 276 $km^2$, and 1 000 ponds of a total of 15 500 ha. The two major rivers of Turkey, Euphrates and Tigris, have an annual discharge of 31.6 $km^3$ and 21.3 $km^3$, respectively. All rivers of Turkey have an annual mean water discharge of 133.2 km³. At present, 193 dams serve environmental protection and flood control, and 151 dams serve irrigation. Water potential by storages is shown in Fig. 1.

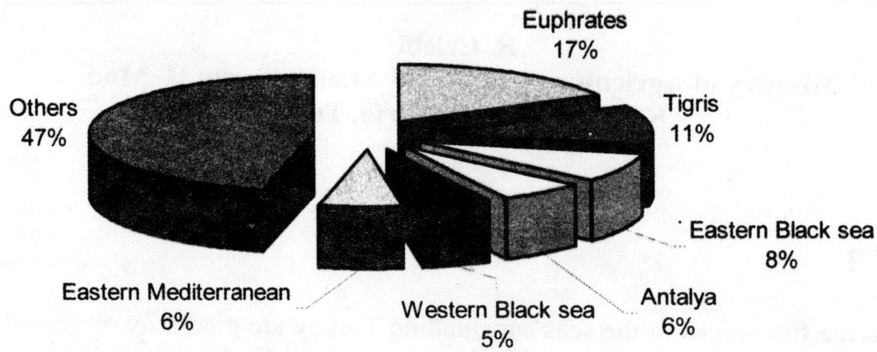


**Fig. 1** Water potential by storages in Turkey

## 2. THE SOUTHEASTERN ANATOLIAN PROJECT (GAP)

In terms of potential, 8.5 million ha of land can be economically irrigated. The current major irrigation project under way is the GAP (Guneydogu Anadolu Projesi), which will eventually result in the irrigation of 75 000 km². GAP is a multi-purpose river basin project using the water of the Euphrates and Tigris and involves the construction of 22 dams. When completed, the GAP will foster investment activities in south-eastern Turkey. The GAP project covers entirely or partially eight provinces. It is more than a mere agricultural project as it includes industry, transportation, tourism, telecommunications, health care, education and rural and urban housing infrastructure projects. It will enhance the region's living standard and income levels, increase productivity and employment opportunities, minimize emigration from the southeast and contribute to the national export potential (Ozaslan, 1998).

### 2.1 Agriculture and water potential of GAP

It is estimated that irrigating 1.7 million ha of land will require construction of many dams of which those constructed already have a total water surface area of 129 987 ha (Table 1). The final reservoir water surface area will reach eventually 198 473 ha. There will also be 2 235 km of irrigation canals. Nineteen dams will serve several purposes, such as irrigation and hydropower production (Dursun, 1999). The Euphrates reservoirs will have a total water surface area of 125 175 ha, and those on the Tigris 73 298 ha. The total reservoir surface area will be 198 473 ha. The GAP project will result in an 11 percent increase in agriculture production, and more than double the personal income. It will provide employment for 3.8 million people, and also produce 6.99 billion kWh of hydropower annually (Anon., 1999).

**Table 1**
Reservoirs in the GAP project area

| River | Operating (ha) | Under construction (ha) | Planned (ha) | Total (ha) |
|---|---|---|---|---|
| Euphrates | 113 308 | 5 916 | 5 951 | 125 175 |
| Tigris | 16 679 | – | 56 519 | 73 298 |
| Total | 129 987 | 5 916 | 62 570 | 198 473 |

Of the 7.54 million ha of the GAP land, 43.6 percent is suitable for agriculture, 29.4 percent for pastures, and 19.2 percent for forestry. This represents 11.7 percent of the agricultural area, 10.3 percent of the pastures, and 6.2 percent of the forests of Turkey. By 1998, irrigated areas in GAP already reached 299 554 ha, of which 193 697 ha were irrigated by state and 105 587 ha by farmers. The major crops produced are wheat, beans, soybeans, nuts, sunflowers, sesame and rice. The region's climate is suitable for double cropping.

Irrigation systems

In the GAP irrigation water is diverted from reservoirs by canals, from which 95 percent of land is irrigated. The remaining 5 percent of land is irrigated using other irrigation methods. The irrigation has already had some side effects, such as salinization of about 20 000 ha of land and rising ground water levels. It is expected that the situation will not get better unless precautions are taken.

Freshwater fish production in the GAP and in Turkey

The GAP represents almost 17 percent of the total inland water resources potential, but the region has a low fish consumption and low fish production. In 1998, capture fisheries harvested 1 842 tonnes, and aquaculture produced 290 tonnes in the GAP region (Table 2). This was equal to 3.3 percent of capture from inland waters of Turkey, and to 1.5 percent of freshwater aquaculture (Celikkale *et al.*, 1999). Carp (*Cyprinus carpio*) dominated the fish production, with 1 200 tonnes, followed by *Capoeta trutta* (320 tonnes), and *Leuciscus cephalus* (78 tonnes).

**Table 2**
Freshwater fish production (in tonnes) in the GAP project area and in Turkey

| Year | GAP region | | | Turkey | | |
|---|---|---|---|---|---|---|
| | Capture | Culture | Total | Capture | Culture | Total |
| 1980 | 805 | - | 805 | 32 225 | - | 32 225 |
| 1985 | 988 | - | 988 | 45 471 | - | 45 471 |
| 1990 | 393 | 10 | 403 | 37 315 | 4 237 | 41 552 |
| 1995 | 1 276 | 125 | 1 401 | 44 983 | 13 113 | 58 096 |
| 1998 | 1 842 | 290 | 2 132 | 54 500 | 33 290 | 87 790 |

The Ataturk reservoir is one of the newest reservoirs constructed. A list of its fish species is given as Table 3. The water quality for Ataturk reservoir is shown in Table 4.

**Table 3**
List of fish species in Ataturk reservoir (Yuksel *et al.,* 1998)

| Family | Species |
|---|---|
| Bagridae | *Mystus halepensis* |
| Poecilidae | *Gambusia affinis* |
| Mugilidae | *Mugil abu* |
| Mastacembelidae | *Mastacembelus simack* |
| Cobitididae | *Nemacheilus (Paragobitis) tigris* |
| Siluridae | *Silurus (Parasilurus) triostegus* |
| Cyprinidae | *Carassius carassius*<br>*Chalcalburnus Acanthalburnus microlepis*<br>*Acanthobrama marmit*<br>*Acanthobrama terrae-sanctae*<br>*Aspius aspius*<br>*Aspius vorax*<br>*Barbus rajanorum mystaceus*<br>*Barbus xanthopterus*<br>*Capoeta capoeta umbla*<br>*Capoeta trutta*<br>*Carassobarbus luteus*<br>*C. mossulensis*<br>*Chondrostoma regium*<br>*Cyprinion macrostomus macrostomus*<br>*Cyprinus carpio*<br>*Garra rufa obtusa*<br>*Garra (Discognatus) variabilis*<br>*Leuciscus cephalus orientalis*<br>*Leuciscus lepidus*<br>*Tor grypus* |

At present there are eight commercial trout cage-culture farms. These farms buy fingerlings (weight 50 to 100 g) from other farms for ongrowing to marketable size within 5-6 months. The culture season extends from mid-October to mid-May. There are two hatcheries in the GAP region. One of them is located near the Ataturk reservoir and belongs to the State Hydraulic Works (DSI). Another is at the Keban dam and also belongs to DSI. Common carp is the other major cultured fish species. Both hatcheries produce fingerlings for stocking the reservoirs as well as other waterbodies. Table 5 shows the number stocked in the GAP reservoirs and the Tigris River in 1999.

## 3. DISCUSSION

Using irrigation, Turkey has increased the crop and other agricultural production by about 100 percent. But the agricultural areas, even though irrigated, are still relatively underutilized, as irrigation practices are relatively new to farmers. There is a need for training and education of farmers and managers, especially in the use of irrigation techniques. The training should include subjects such as double cropping, planting and plant protection, fertilization, soil analysis, water husbandry, including the dangers from rising groundwater levels and salinization, integrated pest management, etc.

**Table 4**

Water quality in Ataturk reservoir (Yuksel *et al.*, 1998)

| Parameter | Value | Year | Jan | Feb | Mar | Apr | May | Jun | July | Aug | Sep | Oct | Nov | Dec |
|---|---|---|---|---|---|---|---|---|---|---|---|---|---|---|
| Temperature | °C | 1995 | 12.0 | 12.0 | 12.0 | 16.0 | 25.0 | 25.0 | 25.0 | 28.0 | 26.0 | 22.0 | 17.0 | 14.0 |
| | | 1996 | 11.0 | 11.0 | 11.5 | 15.0 | 21.0 | 26.0 | 28.0 | 27.5 | 23.5 | 19.0 | 15.0 | 16.0 |
| Conductivity | μmhos $cm^{-1}$ | 1995 | 332 | 330 | 340 | 332 | 362 | 354 | 349 | 340 | 337 | 310 | 364 | 374 |
| | | 1996 | 381 | 360 | 357 | 370 | 365 | 330 | 305 | 338 | 322 | 326 | 347 | 362 |
| Dissolved oxygen | mg/litre | 1995 | 9.9 | 9.8 | 8.6 | 7.9 | 7.0 | 6.9 | 7.1 | 7.4 | 8.5 | 8.4 | 9.0 | 9.5 |
| | | 1996 | 9.9 | 9.6 | 9.6 | 9.9 | 7.3 | 7.1 | 7.1 | 7.5 | 8.0 | 8.3 | 8.8 | 9.0 |
| pH | | 1995 | 6.75 | 7.22 | 7.32 | 7.15 | 7.55 | 7.53 | 7.56 | 8.50 | 8.48 | 8.68 | 8.75 | 8.45 |
| | | 1996 | 8.31 | 8.10 | 8.27 | 8.48 | 8.50 | 7.60 | 7.88 | 7.96 | 8.22 | 8.25 | 8.45 | 7.00 |
| Na | mg/ litre | 1995 | 12.1 | 12.7 | 14.2 | 13.2 | 13.2 | 12.8 | 11.9 | 11.7 | 11.7 | 10.0 | 10.6 | 9.33 |
| | | 1996 | 14.7 | 13.2 | 13.4 | 11.2 | 10.8 | 9.52 | 6.47 | 7.8 | 11.7 | 16.1 | 12.9 | 13.4 |
| K | mg/litre | 1995 | 1.39 | 1.35 | 1.33 | 1.19 | 1.14 | 0.88 | 0.20 | 0.24 | 0.21 | 0.21 | 0.21 | 0.21 |
| | | 1996 | 1.60 | 1.47 | 1.52 | 1.73 | 1.21 | 1.21 | 1.30 | 0.91 | 1.27 | 3.12 | 2.01 | 1.72 |
| $NO_3$ | mg/litre | 1995 | - | 1.71 | - | 2.20 | 1.77 | - | - | - | - | - | - | 2.65 |
| | | 1996 | 1.21 | 1.08 | - | 2.60 | 1.04 | - | - | - | - | - | - | 1.47 |
| $NO_4$ | mg/litre | 1995 | 0.01 | 0.026 | 0.41 | 0.19 | 0.09 | - | 0.24 | 0.05 | 0.12 | 0.75 | 0.75 | - |
| | | 1996 | 0.22 | 0.09 | - | - | - | - | 0.07 | - | 0.017 | - | - | - |
| Phosphate | mg/litre | 1995 | - | - | - | 0.004 | - | 0.006 | 0.006 | 0.016 | 0.003 | - | - | - |
| | | 1996 | - | - | - | - | - | - | - | - | - | - | - | - |
| Salinity | % | 1995 | 0.03 | 0.03 | 0.03 | 0.03 | 0.03 | 0.03 | 0.03 | 0.03 | 0.03 | 0.03 | 0.03 | 0.03 |
| | | 1996 | 0.03 | 0.03 | 0.03 | 0.03 | 0.03 | 0.03 | 0.03 | 0.03 | 0.03 | 0.03 | 0.03 | 0.03 |
| BOD | mg/litre | 1995 | 0.7 | 0.8 | 0.9 | 0.9 | 0.7 | 0.6 | 0.6 | 0.4 | 0.2 | 0.5 | 0.2 | 0.6 |
| | | 1996 | 1.4 | 1.2 | 1.3 | 1.1 | 0.9 | 1.1 | 1.3 | 1.2 | 0.9 | 1.0 | 0.7 | 0.7 |

Regarding fisheries, the existing fish stocks need better protection and rational management, and the local fishermen/farmers should be trained in capture fisheries methods and in aquaculture practices suitable for application in waterbodies of irrigation systems. The existing fishery regulations need to be adapted to the GAP region, which differs climatically from other regions of Turkey. This concerns especially closed seasons for individual fish species. Dam managers responsible for water releases should be made aware of water requirements for fish spawning periods. There is also a need to gradually replace the predominant low-value species by higher value ones in some waterbodies. Better organization of marketing is also required. A cooperative approach to problem solving may be suitable for the integration of capture, processing and marketing of fish.

Aquaculture technologies, such as cage culture, should be more widely applied, as well as testing new fish species for their suitability for the new artificial waterbodies.

Finally, a better administrative support is needed, if the waterbodies serving irrigation in the GAP region are to be fully exploited for their maximum fish production.

**Table 5**
Fish stocked in 1999 in the GAP reservoirs and the Tigris River (Safak, 2000)

| Reservoir/River | Fingerlings (x1000) | Species |
|---|---|---|
| Ataturk reservoir | 15 500 | Common carp |
| Karakaya reservoir | 2 800<br>650 | Common carp<br>Rainbow trout |
| Devegeçidi reservoir | 1 630 | Common carp |
| Hancağız reservoir | 680 | Common carp |
| Hacı Hıdır reservoir | 800 | Common carp |
| Dumluca reservoir | 980 | Common carp |
| Göksu reservoir | 960 | Common carp |
| Çamgazi reservoir | 325 | Common carp |
| Kıral Kızı reservoir | 500 | Common carp |
| Tigris river | 200 | Common carp |
| Batman reservoir | 100 | Common carp |
| Total | 24 475<br>650 | Common carp<br>Rainbow trout |

## REFERENCES


Anon., 1999. *Dams and Hydroelectric Power Plants in Turkey*. Ministry of Energy and Natural Resources. General Directorate of State Hydraulic Works, Ankara.

Celikkale, M.S., Duzgunes, E. & Okumus, O. 1999. Turkie su urunleri sektoru, potansiyeli, mevcut durumu, sorunlari ve cozum onerileri. Istanbul Ticaret Odasi Yasin No. 1999-2, Istanbul.

Dursun, H. 1999. Turkiye'nim toprak kaynaklari ve kullanimi. Tarimda su kullanimi ve yonetimi sempozyumu. Turk Yuksek Ziraat Muhendisleri Birligi ve Vakfi, Ankara.

Ozaslan, I. 1998. Natural Resources of Turkey. Vol. 1. Istanbul Chamber of Commerce. Publication No. 1998-77, Istanbul.

Safak, N. 2000. Bolgesi'ndeki baraj gollerinde DSI Genel Mudurlugunce Yurutulen ve Planlanan Su Urunleri Calismalari. DSI Genel Mudurlugu, Ankara.

Yuksel, M., Sevik, R. Yapalak, S. & Hartavi, S.M. 1998. Ataturk baraj golu'nun su kalitesi ve bu sulara uyum gosteren balik turlerinin tesbiti. GAP 2. Muhendislik Kongresi, 21-23 Mayis 1998, Sanliurfa.

# THE USE OF IRRIGATION SYSTEMS FOR SUSTAINABLE FISH PRODUCTION IN INDIA

by

**B.P. Das**
**717, Saheed Nagar, Bhubaneshwar, India, 751007**

## ABSTRACT

India, with arable land of 169.70 million hectares (mha), has large irrigation systems commanding a gross area of almost 90 mha. India has the distinction of being the country with the largest irrigation coverage in the world. Per capita availability of land and fresh water is shrinking for the large population of 1 billion. Because of erratic distribution of monsoon rainfall (June through October), varying from 200 mm in the arid west to 10 000 mm in the mountainous northeast, provision of irrigation has been given a thrust to sustain agriculture. Irrigation coverage has expanded from 21 mha in 1951 to 90 mha in 2000. A major component of the irrigation sector is large (generally multipurpose), medium and minor reservoirs. These reservoirs have a combined volume of 250 billion cubic metres (BCM) and water area of around 40 000 km². The largest man-made lake in southern Asia is Hirakud reservoir (740 km²) built (1957) on the Mahanadi River in Orissa. Very large canals carrying up to 1 000 m³/sec, such as the Rajasthan canal, form another major component of irrigation systems. These man-made waterbodies have large fish production potential. Annual fish yields in large Indian reservoirs range from 0.5 to 109.2 kg/ha. Smaller medium and minor reservoirs have yields of up to 300 kg/ha, with enhancement. Overall annual production from all inland waterbodies in India is at present 2.5 million tonnes against the potential of at least 10 million tonnes. In a large canal with cages and in smaller outlet channels a low-through culture system trials yielded 100 tonnes/ha/year. Stocking, appropriate water release schedule, fishing gear regulation, appropriate management of the upstream zone, maintaining migratory path, self reliance and people's participation have been identified as beneficial to fisheries.

## 1. WATER RESOURCES OF INDIA AND IRRIGATION DEVELOPMENT

India, which covers an area of 329 mha, is a country endowed with vast water and land resources. With arable land of 169.70 mha, it ranks third to Russia's 208.4 mha and the United States of America's 187.76 mha. With its population of one billion accounting for one sixth of the world population, its renewable water resources are only 4 percent of global availability. Per capita availability of land and fresh water in India is 0.2 ha and 220 m³ against the world average of 0.27 ha and 7 400 m³, respectively.

Indian economy has been based on traditional agriculture for centuries. The net cropped area of 140 mha experiences erratic monsoon both spatially and temporally (June through October), with precipitation ranging from a meager 200 mm in the western Rajasthan desert to 10 000 mm in the northeastern Himalayan states. India frequently experiences drought and floods that seriously affect crop yields. Since independence (1947), in order to sustain agriculture, water resources development projects with major objectives of irrigation, flood control and power generation have been implemented to progressively harness the annual precipitation of 4 000 billion cubic metres (BCM) of which the utilizable quantity is 1 122 BCM as shown below.

| Classification | Quantity (in BCM) |
|---|---|
| Surface water | 690 |
| Ground water | 432 |
| Total | 1122 |

Between 1951 and 2000 the population of India grew from 361 million to a billion. Food production has also caught up, with an increase from 51 million tonnes in 1951 to 200 million tonnes in 1999, primarily attributable to the implementation of irrigation schemes comprising of 311 major projects, 944 medium projects and 40 000 minor projects.

India has achieved the gigantic task of harnessing its water for managing and sustaining agriculture primarily through large flow irrigation projects having major, medium and minor reservoirs as head works. While surface water irrigates almost 46 mha, ground water irrigates another 46 mha of arable land.

Over the last 50 years there has been a steady increase in irrigated lands (Table 1). Additional 6 mha is likely to be irrigated by 2002 during the 9th plan.

**Table 1**

Chronological increase in irrigated lands (in million hectares)

| PLAN | Potential created in mha | | | | |
|---|---|---|---|---|---|
| | Major & medium | Minor | | | Total |
| | | Surface water | Ground water | Total | |
| Pre-plan to 1951 | 9.70 | 6.40 | 6.50 | 12.90 | 22.60 |
| I Plan (1951–56) | 12.20 | 6.43 | 7.63 | 14.06 | 26.26 |
| II Plan (1956–61) | 14.33 | 6.45 | 8.30 | 14.75 | 27.08 |
| III Plan (1961–66) | 16.57 | 6.48 | 10.52 | 17.00 | 33.57 |
| Annual Plan (1966–69) | 18.10 | 6.50 | 12.50 | 19.00 | 37.10 |
| IV Plan (1969–74) | 20.70 | 7.00 | 16.50 | 23.50 | 44.20 |
| V Plan (1974–78) | 24.72 | 7.50 | 19.80 | 27.30 | 52.02 |
| Annual Plan (1978–80) | 26.61 | 8.00 | 22.00 | 30.00 | 56.61 |
| VI Plan (1980–85) | 27.70 | 9.70 | 27.82 | 37.52 | 62.22 |
| VII Plan (1985–90) | 29.92 | 10.99 | 35.62 | 46.61 | 76.53 |
| Annual Plan (1990–92) | 30.74 | 11.46 | 38.89 | 50.35 | 81.09 |
| VIII Plan (1992–97) | 33.82 | 12.24 | 45.73 | 57.97 | 91.79 |
| End of IX Plan (likely) 1997–2002 | 38.00 | 18.00 | 47.00 | 60.00 | 98.00 |

An examination of the investment in irrigation development in India through major and medium projects shows that from the irrigation of a gross area of 85 mha, at least 70 million families having a population of 350 million, have benefited (Fig. 1). Currently it costs Rs. 90 000 (US $ 2 000) to bring 1 ha of land into production.

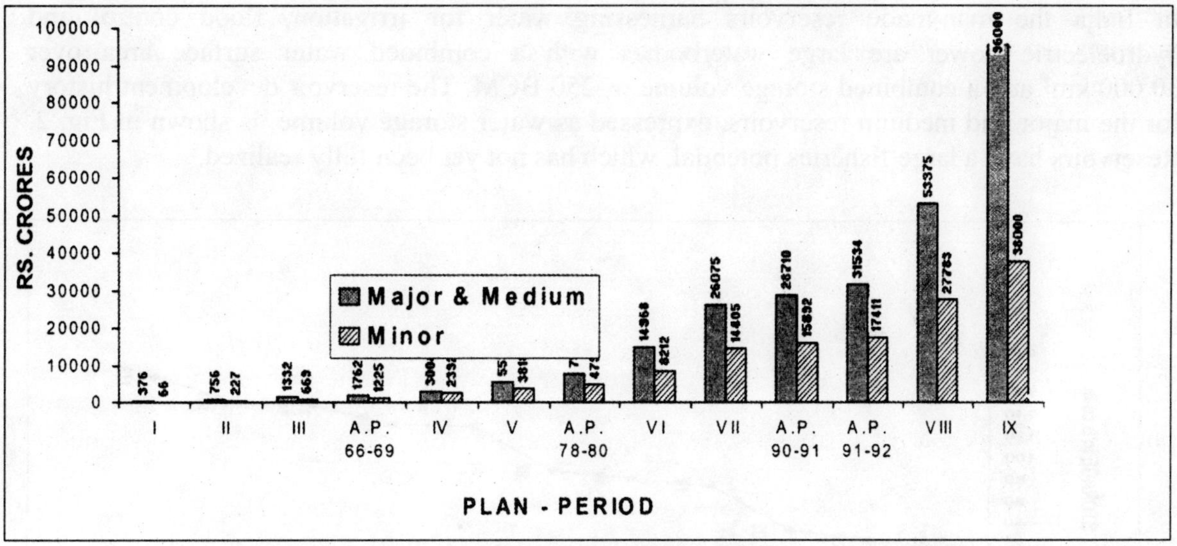


**Fig. 1** Planwise financial expenditure (cumulative) at current prices (Note: crore = 10 million)

It is obvious that the extremely high investment India makes in irrigation development must aim at integrated utilization of the harvested water for sustainable rural development such as horticulture, pasture, aquaculture and capture fisheries, in addition to agriculture. Appropriate investment in the socio-economic fields and for research, extension and training is imperative for the irrigation sector.

## 2. WATER FOR IRRIGATION

Of the gross cropped area of 190 mha in the country, food grain accounts for 126 mha and the rest are cash crops. The gross irrigation coverage is 85 mha/annually. Irrigation in India is supplemental during the Kharif season (June to October) for the principal cereal crops rice, maize, jowar, etc., and almost total during the Rabi season (November to April) for wheat and other cash crops. With 45 mha under rice, of which 23 mha is irrigated, and 27 mha of under-irrigated wheat, the bulk of consumption of the reservoir water is for these crops due to continuous irrigation for four months for each principal crop. Cash crops consume almost 25 percent of the stored water but with intermittent irrigation. The average irrigation supplementation for typical crops, which varies significantly from loamy soil to clayey soil and from low rainfall areas (less than 600 mm annually) to high rainfall areas (around 2 000 mm annually or more), is as follows:

| Crop | Supplementation in mm | |
|---|---|---|
| | Kharif | Rabi |
| Rice | 700 – 1600 | 1600 |
| Maize | 100 – 300 | 500 – 600 |
| Groundnut | 300 – 400 | 500 –700 |
| Cotton | 150 – 300 | 600 – 800 |
| Sugar cane (perennial) | 600 –800 | |

In India the man-made reservoirs harnessing water for irrigation, flood control and hydroelectric power are large waterbodies with a combined water surface area over 30 000 km² and a combined storage volume of 250 BCM. The reservoir development history for the major and medium reservoirs, expressed as water storage volume, is shown in Fig. 2. Reservoirs have a large fisheries potential, which has not yet been fully realized.

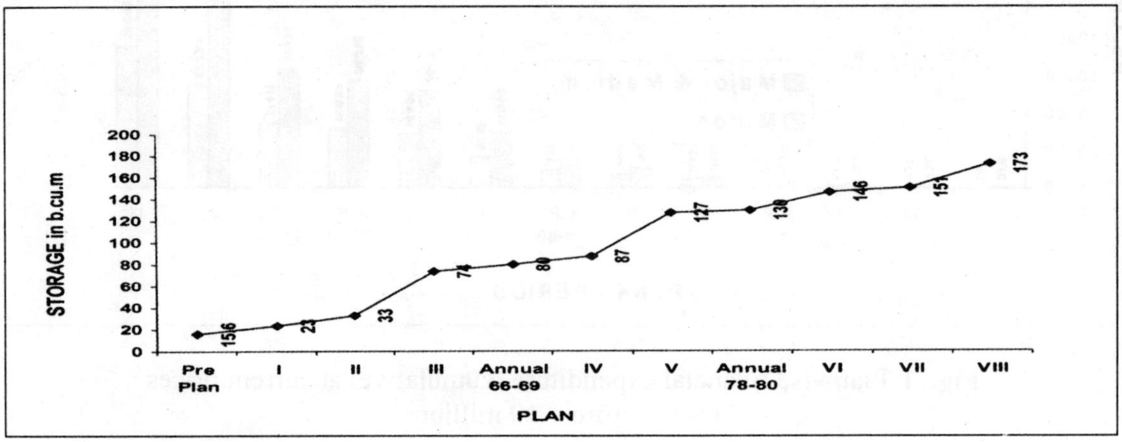


**Fig. 2** Creation of storage during Plan Development

## 3. SOME PROBLEMS

In an irrigated area the pattern of drawing irrigation water is not only related to the crop water needs, but also has to respond to a large number of other water users. During droughts when water management is not well formulated, interference with operation, unauthorized water diversion and improper irrigation practices such as ponding, are common. These factors can be redressed by social involvement of a group of farmers through Participatory Irrigation Management (PIM), which is proving to be a great success in economizing water use and concurrent maintenance of the health of the system. Institutional strengthening incorporating beneficiary participation is therefore a priority.

Excessive water withdrawal leads to accumulation of surplus water over lowlands, causing waterlogging and also increase in the ground water table. It is recognized that of the 255 mha of area irrigated worldwide one third is threatened by waterlogging and salinity. Almost 3 mha out of 65 mha of net irrigated land in India is already suffering from waterlogging. For irrigation to be efficient, participatory and integrated management of irrigation and drainage is essential.,

## 4. USE OF IRRIGATION SYSTEMS FOR FISH PRODUCTION

Currently 38 mha is irrigated through 1 300 major and medium Flow Projects and 13 mha through minor Flow Projects. This includes almost 25 mha of arid/semi-arid lands. These projects essentially have a reservoir whose area varies from less than 1 $km^2$ to 740 $km^2$. The Hirakud project (Fig. 3), with a largest reservoir in India covering 740 $km^2$, has a large network of canals and provides irrigation to 0.16 mha. The main canal (130 $m^3$/sec) of up to

50 m bed width is 95 km long, feeding almost 3 000 km of secondary canals of up to 20 m bed width.

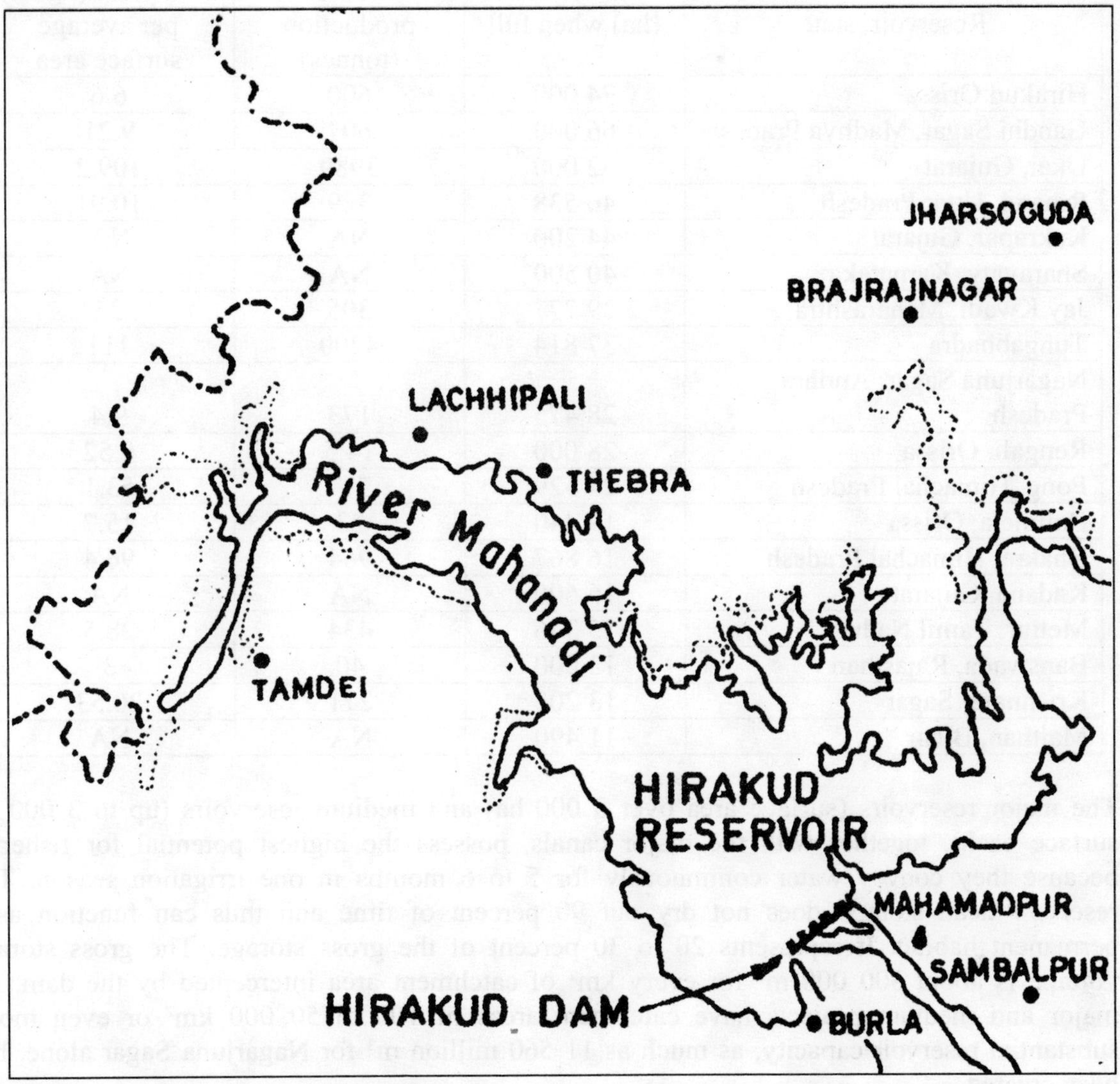


**Fig. 3** Hirakud reservoir, Orissa

Apart from irrigation from reservoirs, eastern and southern India have large delta irrigation systems fed by diversion structures such as those on the rivers Mahanadi, Godavari, Krishna and Cauvery, and the rivers Kosi, Sone, Gandak, Damodar in Bihar and West Bengal, and the Ganga and Yamuna canals in the alluvial plains of Uttar Pradesh. These long anicuts/barrages up to 4 km long support large but shallow reservoirs of 3 to 5 m depth, with areas of up to 2 000 ha, which provide a continuous water supply to canals. Fish yields in these shallow reservoirs are estimated to range from 5 to 10 kg/ha. Fish production could possibly be higher if fish were allowed to migrate in and out of these reservoirs and if quick releases of water were avoided. The largest irrigation canal of India currently operating is the Rajasthan canal which receives water from a reservoir stored behind the 255 m high Bhakra dam. The reservoir supplies water for irrigation of 4 mha.

**Table 2**

Surface area of some major reservoirs in India and their fish production

| Reservoir, state | Surface area (ha) when full | Annual production (tonnes) | Fish yield (kg/ha) per average surface area |
|---|---|---|---|
| Hirakud,Orissa | 74 000 | 600 | 6.6 |
| Gandhi Sagar, Madhya Pradesh | 66 000 | 607 | 9.21 |
| Ukai, Gujarat | 52 000 | 3989 | 109.2 |
| Rihand, Uttar Pradesh | 46 538 | 329 | 10.91 |
| Kakrapar, Gujarat | 44 200 | NA | NA |
| Sharavaty, Karnataka | 40 500 | NA | NA |
| Jay Kwadi, Maharashtra | 39 777 | 305 | 25 |
| Tungabhadra | 37 814 | 4200 | 111 |
| Nagarjuna Sagar, Andhra Pradesh | 28 475 | 173 | 9.4 |
| Rengali, Orissa | 28 000 | 14.5 | 0.52 |
| Pong, Himachal Pradesh | 24 529 | 797 | 53.1 |
| Balimela, Orissa | 19 440 | 23 | 15.7 |
| Bhakra, Himachal Pradesh | 16 867 | 964 | 96.4 |
| Kadana, Gujarat | 16 600 | NA | NA |
| Mettur, Tamil Nadu | 15 346 | 434 | 28.5 |
| Banswada, Rajasthan | 13 500 | 40 | 3 |
| Krishnaraj Sagar | 13 200 | 271 | 20.53 |
| Maithan, Bihar | 11 490 | NA | NA |

The major reservoirs (surface area over 3 000 ha) and medium reservoirs (up to 3 000 ha surface each), together with the major canals, possess the highest potential for fisheries because they convey water continuously for 5 to 6 months in one irrigation season. The reservoir dead storage does not dry out 90 percent of time and thus can function as a permanent habitat. It represents 20 to 30 percent of the gross storage. The gross storage volume is about 300 000 $m^3$ for every $km^2$ of catchment area intercepted by the dam. As major and medium projects have catchment areas of 100 to 50 000 $km^2$ or even more, substantial reservoir capacity, as much as 11 560 million $m^3$ for Nagarjuna Sagar alone, has been created.

Shallow and flat Indian reservoirs have a higher fish yield than deep and elongated reservoirs. The operational trend for a reservoir is to go from the dead storage level (DSL) in June to full reservoir level in October, followed by water depletion. This profoundly influences fish production. Salandi reservoir in Orissa, with 3 100 ha of surface area, depletes quickly in drought years and this results in lower fish yields, i.e. about 10 kg/ha against 20 kg/ha in normal years.

Simulation of the past hydrological data, including topography and deposition of sediments, may assist in the preparation of a rational plan for maintaining optimal sustainable level of fishing effort. In smaller reservoirs appropriate stocking in June-July and harvesting in April-May have given good results, but this is yet to be achieved for large reservoirs.

Canals also have a great potential for fisheries. Limited data on cage culture under controlled conditions between cross-regulators in some large canals such as in the Mahanadi delta system with the main canal of 45 m bed width, and Hirakud canal of 40 m bed width, have shown promise for cost-effective aquaculture, provided that a people's participation programme is functional.

The institutional arrangement for obtaining optimal fisheries production needs attention. Water resources organizations which build and operate reservoirs focus primarily on meeting the irrigation or other needs without an established or formulated linkage to fisheries. In India every major reservoir is operated by following a "rule curve", which maintains certain levels at a certain period in the year, which is linked with the hydrology of the basin and consumptive/non-consumptive requirements. As soon as a reservoir in India becomes operational, fisheries is generally initiated by a public sector corporation. Alternately the reservoir fishing rights are leased to a fishermen cooperative society at a particular royalty rate. Generally there are two types of cooperative societies, viz. the primary cooperative engaged in fishing, and the apex society engaged in marketing and providing marketing infrastructure and logistics. The fishery development scheme applying stock enhancement, tried in Malampuzha Reservoir in Kerala (Drewes, 1994) through an Indo-German fisheries development project, seems appropriate for Indian conditions (Natarajan *et al.*, 1979).

## 5. IMPACT OF IRRIGATION DEVELOPMENT ON FISHERIES

India's inland water area has kept increasing with the completion of new irrigation reservoirs. The two large reservoirs in India prior to 1956, when dam construction was accelerated, were Mettur in Tamil Nadu, with water surface area of 15 346 ha, and Tungabhadra in Karnataka, with a surface of 37 814 ha. Fisheries production in Tungabhadra increased from 15 tonnes in 1958 to 156 tonnes in 1965, and to 4 200 tonnes in 1981–82. Between 1980 to 1987, it was fluctuating between 2 000 and 1 800 tonnes/year and had fallen to 1 600 tonnes in 1993. The large reservoirs which appeared since 1957, such as Hirakud, Bhakra, Nargarjuna Sagar, Pong, Rihand and Ukai have average yields of 10 to 60 kg/ha, whereas in the 500 medium reservoirs the yields have been 20 to 30 kg/ha.

With more than 3 million ha of reservoirs in India, freshwater fish production had gone up from 0.22 million tonnes in 1951–52 to 1.6 million tonnes 1990–91 and to 2.5 million tonnes in 1997.

Due to inherent logistical advantage the small and medium reservoirs show consistently higher yields than large reservoirs.

The fish production potential of all freshwaterbodies of India is estimated to be at least 10 million tonnes provided appropriate and scientific management is practised.

A firm institutional linkage between the Water Resources Department and the Fisheries Department in the states of India has not yet been established and information on fisheries practices is not yet available for all reservoirs.

**Table 3**

Freshwater fish production in India in million tonnes

| Year | Inland fish production |
|---|---|
| 1950–51 | 0.218 |
| 1960–61 | 0.280 |
| 1970–71 | 0.670 |
| 1980–81 | 0.887 |
| 1990–91 | 1.536 |
| 1997–98 | 2.438 |

Fisheries scientists and managers need to be acquainted with the average pattern of irrigation water demand so as to adjust their stocking and harvesting programmes accordingly. Rapid fluctuation of reservoir water level, which is common especially in irrigation reservoirs, has been the primary cause of low yields (Jhingran, 1988). Detailed evaluation of this impact is urgently required. For each reservoir maps of shoreline at different water levels need to be made available to assist in the decision on positioning of enhancement facilities, such as cages and pens. On large canals there is a need for fish culture pilot trials. Some information is already available from the canal fish production studies conducted in Orissa (Anon., 1991; Saha and Paul, 2000).

## 6. POTENTIAL FOR THE USE OF IRRIGATION SYSTEM FOR FISH PRODUCTION

Any irrigation system has two major components: the headworks, and the distribution network. Reservoirs forming the headworks constitute the major habitat for fish. The large and medium reservoirs do not dry up which is a major benefit from a conservation viewpoint. Reservoir management using fish stock enhancement results in significant increases in fish yield (Table 4).

Table 5 shows the extreme variability in fisheries production among Indian states: West Bengal with 865 700 tonnes, as compared with 70 330 tonnes in Gujarat.

**Table 4**

Reservoirs showing the enhancement impact on fish yields

| Years | Average yield kg/ha/year | | |
|---|---|---|---|
| | Kerala | Bhakra | Ukai |
| 1978–79 | 6.60 | 75 | 21 |
| 1987–88 | 6.70 | 38 | 79 |
| 1988–89 | 15.10 | 77 | 109 |
| 1989–90 | 17.50 | 81 | |
| 1990–91 | 20.40 | 80 | |
| 1991–92 | 25.50 | 85 | |
| 1992–93 | 23.38 | 95 | |

**Table 5**
Fisheries production by state

| State | Annual freshwater fish production in thousand tonnes | Waterbodies ha x 105 | Reservoirs ha x 105 |
|---|---|---|---|
| West Bengal | 865.70 | 5.45 | NA |
| Bihar | 254.74 (not confirmed) | 1.60 | 0.97 |
| Uttar Pradesh | 192.91 | 4.45 | 3.35 |
| Assam | 159.77 | 1.35 | NA |
| Maharashtra | 135.39 | 3.39 | 2.74 |
| Orissa | 135.30 | 9.67 | 2.56 |
| Madhya Pradesh | 127.43 | 4.13 | 4.60 |
| Karnataka | 126.64 | 6.42 | 4.38 |
| Kerala | 73.90 | 3.03 | NA |
| Gujarat | 70.33 | 7.02 | 2.87 |

## 7. CONSTRAINTS AND REMEDIES

Fisheries production benefits from the following:

- stocking and stock enhancement; annual evaluation;
- self reliance; fishing gear mesh size regulation;
- nutrient enrichment during floods;
- appropriate management in the reservoir shallow upper zone with lotic environment which serves as breeding and nursery ground;
- the middle and lower lacustrine zone should be developed for year-round commercial fishing; in the early part of monsoon, with the reservoir at the lowest drawdown level, flood water enters the reservoir at a higher current velocity of up to 3 m/sec, and dam spilling may wash out the fry and fingerlings, as observed for example in Konar reservoir in Bihar (Parameshwaran, Khan and Ramakrishnaya, 1978). Quick filling up of a reservoir during the first week of July (without any spilling) will reduce this hazard to fish stocks.

Fish production may face the following constraints:

- aquatic macrophytes in the reservoir, which retard increase in stocks of some fish species. Submerged weeds also provide shelter for some weed fish;
- decomposition of large quantities of aquatic plants will result in reduced dissolved oxygen levels or complete deoxygenation;
- presence of dead trees which were not clear-felled prior to flooding, obstructs the use of fishing gear.

When a dam is constructed, riverine fish species suffer from the loss of their habitat due to changes in water current and turbidity, loss of breeding grounds and changes in fish food organisms.

A dam also represents an obstruction to fish migration. Very few dams have fish ladders, and those which do are mostly unsuitable for the migration of the Indian fish species. An example is the dammed Cauvery River where, after the construction of anicuts in the coastal region, the upriver fish migration was totally blocked.

## 8. ORISSA AND REGIONAL EXPERIENCE

Among the states of India Orissa has the largest waterbody area of almost one million hectares. This offers a large scope for fisheries in reservoirs, rivers and canals. Reservoirs cover an area of 256 000 ha, the largest component of all waterbodies in this state.

Eastern, southern and southwestern India, except the states Rajasthan, Haryana, Punjab, Jammu-Kashmir, get a high (up to 1 500 mm) monsoon rainfall, which results in substantial filling (3 out of 4 years) of the irrigation reservoirs, 75 percent of the year. Surface areas of these reservoirs (Sugunan, 1995) represent almost 90 percent area of all Indian reservoirs.

Reservoirs in Orissa have lower fish yields than rivers, canals, ponds and tanks (Table 6). The main reason is probably the low stocking rate which, according to the recommendation of the Fish Seed Committee of the Government of India (1966), should be 500 fingerlings of 40 to 150 mm size per ha. As in other states of India, the medium and minor reservoirs with enhanced fisheries show considerably higher yields than large reservoirs with poor rates of enhancement.

The Mandira reservoir on the Sankh River is exclusively meant to provide water to a large steel plant with a steady water supply of 5 to 10 m³/sec. The reservoir fills up in July, receiving water from a forested catchment of 3 000 km². The reservoir remains full until December, with occasional spill of perhaps 10 days during the monsoon. The reservoir provides good environment for fish. The stocking rate is 300 to 400 fingerlings per hectare. With a ready market of 300 000 people in the nearby steel city, there is a good tie-up between marketing and producer.

**Table 6**

Waterbodies in Orissa

| Year | Reservoirs (256 000 ha) | | Rivers & canals (155 420 ha) | | Ponds & tanks (14 822 ha) | |
|---|---|---|---|---|---|---|
| | Production (tonnes) | Yield (kg/ha) | Production (tonnes) | Yield (kg/ha) | Production (tonnes) | Yield (kg/ha) |
| 1995–96 | 12 506 | 49 | 12 509 | 81 | 93 079 | 810 |
| 1996–97 | 14 652 | 57 | 14 556 | 93 | 92 927 | 800 |
| 1997–98 | 14 628 | 56 | 16 664 | 107 | 99 980 | 870 |
| 1998–99 | 16 202 | 63 | 26 318 | 169 | 98 386 | 860 |
| 1999–2000 | 13 809 | 54 | 20 873 | 134 | 88 107 | 767 |

| Reservoir | Surface area (ha) | Yield (kg/ha) |
|---|---|---|
| Hirakud | 74 000 (average 50 000) | 6.6 |
| Mandira | 3 330 | 55.13 |
| Pilasalki | 120 | 152.75 |

Pilasalki reservoir (established in 1976 in Phulbani district), is drought prone. The 7-m deep reservoir spills only 10 to 20 percent of its volume. The reservoir is close to a district town. Good breeding in sheltered inshore areas and stocking have resulted in a high yield of 152.75 kg/ha.

The large Hirakud reservoir has low fish yield (Yadav and Sugunan, 1992), partially due to the live storage capacity being only 15 percent of the annual runoff. In a medium rainfall year the fish production is 700 to 800 tonnes (12 to 15 kg/ha), but drops to 300 tonnes in high rainfall years, as both fingerlings and small fish are washed out over the spillways. Stocking is not very systematic. In shallow parts of the reservoir influenced by the inflowing river the yield reaches 20 kg/ha.

## 9. FLOW-THROUGH SYSTEM

Recent research on a flow-through system, carried out at the Central Institute for Fresh Water Aquaculture at Bhubaneshwar, India, using canal water in the Mahanadi delta irrigation system, gives promising results. In the Kanas Branch Canal System (15 $m^3$/sec head discharge) water was passed through 27 tanks, each of 10 $m^3$ volume constructed on a small off-take channel carrying up to 20 litre/sec. The main advantage of the flow-through system was good water oxygenation. With pelletted food and stocking density of 800 per 10 $m^3$, the fish production was 15 kg/10 $m^3$/three months, which is the time during which the water flow in the canal is maintained in one irrigation season. The highest production achieved was 100 tonnes/ha/year at a stocking density of 500. A prototype trial in a large canal can provide a good yield of 10 tonnes/ha/year even with stocking density of 100 fingerlings per 10 $m^3$, provided the inlet and outlet of the canal are suitably blocked to prevent the escape of fish.

Following the experience in Kyrdenkulai reservoir in N.E. India, cage and pen culture has been tried in a limited way in running canals and in Kolab reservoir. Yields up to 100 kg/ha have been reached using cage culture.

## 10. CONCLUDING REMARKS

Small reservoirs achieve the best fish yields when practicing enhanced capture fisheries. Capture–cum–culture approach is recommended for fisheries in medium reservoirs. Achieving a yield of 100–300 kg/ha from the almost 25 mha of medium and small waterbodies which exist in India is a key management issue.

Large reservoirs of India have high fisheries potential if adequate stocking and good management are practiced. These reservoirs are currently producing only 100 000 tonnes of fish per year, but have an estimated potential of 300 000 tonnes. A difficult issue is pollution from industries and other sources.

Cage and pen fish culture in reservoirs and irrigation canals should be practiced on a much larger scale.

People's participation is a necessary condition for successful fisheries management of the large diversity of waterbodies serving and arising from irrigated agriculture.

## REFERENCES


Anon., 1991. Proc. National Symposium on Freshwater Aquaculture. News Item, CIFA. Bhubaneshwar, India.

Drewes, E. 1994. Reservoir fisheries management by co-operatives: issues and scope. Indo-German Reservoir Fisheries Development Project PN: 882134 0-01.200. Malampuzha, Kerala, India.

Jhingran, A.G. 1988. Reservoir Fisheries Management in India. Bulletin 45, CICFRI. Barrackpur, India. 68p.

Natarajan, A.V. 1979. Planning strategies for development of reservoir fisheries in India. Madras J. Fish. 8: 69–77.

Parameshwaran S., Khan, M.A. & Ramakrishnaya, M. 1978. Pre-recruitment studies in Konar reservoir, Bihar. Proc. Seminar on Ecology and Fisheries of Freshwater reservoirs. CICFRI, Barrackpore.

Saha, C. & Paul, B.N. 2000. Flow through system for industrial aquaculture in India. Aquaculture Asia 5(4): 3.

Sugunan, V.V. 1995. Reservoir Fisheries of India FAO Fisheries Technical Paper No. 345. Rome, FAO. 423p.

Yadav, Y.S. & Sugunan, V.V. 1992. Hirakud reservoir – strategies for fisheries Development. Bulletin 66. CICFRI, Barrackpore, India. 18p.

# IRRIGATION IN INDIA

**by**

**P.V. Dehadrai**
**DIII/3403, Vasant Kunj, New Delhi-110070, India**

## ABSTRACT

India has many rivers whose total catchment area is estimated to be 252.8 million ha (mha) Out of about 1 869 km³ of surface water resources, about 690 km³ of water is available for different uses. The ultimate irrigation potential of the country has been estimated to be 139.5 mha. India has acquired an irrigation potential of about 84.9 mha against the ultimate irrigation potential. About 360 km³ of groundwater is also available for irrigation. For soil and water conservation on agricultural lands, both agronomic and mechanical measures are deployed, such as contour farming, management of soil fertility, water nutrient interrelationship, balanced IPNS system and micronutrients. The paper gives details on irrigation water management for major crops in India: rice, wheat, maize, pulses, pigeon pea, gram, mung bean, black gram, field peas, lentil, oil seed crops, sugar cane and cotton. Irrigation is implemented through various methods such as border irrigation, basin irrigation, sprinklers and drip irrigation.

## 1. INTRODUCTION

Water is the most critical input for enhancing agricultural productivity, and therefore expansion of irrigation has been a key strategy in the development of agriculture in the country. The ultimate irrigation potential of India has been estimated to be 139.5 mha, comprising 58.5 mha from major and medium schemes, 15 mha from minor irrigation schemes and 66 mha from groundwater exploitation. India's irrigation potential has increased from 22.6 mha in 1951 to about 90 mha at the end of 1995. It is estimated that even after achieving the full irrigation potential, nearly 50 percent of the total cultivated area will remain rain fed.

If we analyse agricultural growth during the past four decades, we find that high-yielding varieties, irrigated area expansion and fertilizer use have been the major factors contributing to the achievement of green revolution in India. The present level of consumption of total nutrients (NPK) is 14.3 million tonnes/year. On an all-India basis, per-hectare consumption of fertilizer (NPK), which was a meagre 2.0 kg/ha during the early sixties, has risen tremendously during the last 35 years or so, to a level of 76.5 kg/ha.

The share of water use other than for agriculture was only 13 percent in 1985, which is likely to become 27 percent by 2025. Such a fast growth of water need in the face of emerging supply constraints is likely to result in a wide supply gap for irrigation water in the near future.

## 2. WATER RESOURCES OF INDIA

The annual precipitation including snowfall, which is the main source of water in India, is estimated to be in the order of 4 000 $km^3$. The annual potential evapo-transpiration in the country is 1 775 mm, but it varies from a minimum of 1 239 mm in Jammu and Kashmir to a maximum of 2 052 mm in Andhra Pradesh.

The total catchment area of Indian rivers is estimated to be 252.8 mha. The Ministry of Water Resources, Government of India, for the purpose of efficient water management has divided the whole country into 20 river basins, comprising the major basins and including all the other remaining medium and small river systems. Catchment area and water resource potential of these basins is given in Table 1.

The total surface water resources of the country (yearly average flow rate) are about 1 869 $km^3$. It has been estimated that due to extreme variability in precipitation, which disallows assured storage of all flash and peak flows, and due to non-availability of suitable storage sites in hills and plains, only about 690 $km^3$ water is available for different uses. It is further estimated that about 360 $km^3$ of groundwater would be available for irrigation. Thus a total of 1 050 $km^3$ of utilizable surface and groundwater is available for irrigation. The projected budget of water for the twenty-first century is shown in Table 2.

### 2.1 Waterbodies and storages

Inland water resources of India are classified as rivers, canals, reservoirs, tanks and ponds, beels, oxbows, lakes, derelict water and brackish water, and they cover an area of 7 mha excluding rivers and canals. Uttar Pradesh has the highest total length of rivers and canals of 31 200 km, followed by Jammu & Kashmir and Madhya Pradesh. The total length of navigable waterways in the country is estimated to be 15 783 km, and 83 percent of this is in 10 important rivers of the country.

Most tanks and ponds (2.3 mha) are present in the southern states of Andhra Pradesh, Karnataka and Tamil Nadu. The states Andhra Pradesh, Gujarat, Karnataka, Madhya Pradesh, Orissa, Maharashtra, Rajasthan and Uttar Pradesh account for large reservoirs with a total water surface area of 2.1 mha. Most of the lakes, beels and derelict waterbodies lie in the states of Kerala and Tamil Nadu, whereas Orissa, West Bengal and Kerala have the largest areas of brackish water (Table 3). Overall, the inland water resources are unevenly distributed in the country, and five states, i.e. Orissa, Tamil Nadu, Andhra Pradesh, Karnataka and West Bengal, account for more than 50 percent of the inland waterbodies. Today water storages hold 174 $km^3$ of water, and another 72 $km^3$ of the storage capacity will become available after the completion of the ongoing projects. Another 132 $km^3$ of storage capacity is likely to be available from projects under consideration, making a total storage of 378 $km^3$ as compared with the total water resources, including groundwater, available for irrigation, of 2 229 $km^3$ in the country. Additional storage is available in the basins of Ganges, Krishna, Godavari, Narmada, Mahanadi, Brahmaputra and Barak, where the already existing live storage is also high. Tapi basin is leading in terms of storage capacity.

**Table 1**

**Catchments area, average surface and groundwater potential in individual river basins, India, 1998 (million $m^3$ = $Mm^3$)**

| River basin | Catchments area ($km^3$) potential ($Mm^3$) | Average surface water potential ($Mm^3/km^2$) | Average groundwater potential ($Mm^3/km^2$) | Water resources potential ($Mm^3$) | Groundwater ($Mm^3$) |
|---|---|---|---|---|---|
| 1. Indus | 321 289+ | 0.2282 | 0.0790 | 73 305 | 25 543 |
| 2. a. Ganges | 861 452 | 0.6095 | 0.1993 | 525 023 | 171 725 |
| b. Brahmaputra | 41 723 | 2.7634 | 0.1433 | 537 240 | 27 857 |
| c. Barak & others | 199 413 | 1.1590 | 0.0430 | 48 357 | 1 795 |
| 3. Godavari | 312 812 | 0.3534 | 0.1495 | 110 540 | 46 762 |
| 4. Krishna | 258 948 | 03017 | 0.1029 | 78 124 | 26 646 |
| 5. Cauvery | 81 115 | 0.2632 | 0.1676 | 21 358 | 13 598 |
| 6. Subarnarekha | 29 196 | 0.4236 | 0.0748 | 12 368 | 2 185 |
| 7. Brahmani-Baitarani | 51 822 | 0.5495 | 0.1134 | 28 477 | 58 790 |
| 8. Mahanadi | 141 589 | 0.4723 | 0.1504 | 66 879 | 21 293 |
| 9. Pennar | 55 213 | 0.1144 | 0.0914 | 6 316 | 5 047 |
| 10. Mahi | 34 842 | 0.3163 | 0.1399 | 11 020 | 4 875 |
| 11. Sabarmati | 21 674 | 0.1757 | 0.1399 | 3 809 | 3 033 |
| 12. Narmada | 98 796 | 0.4620 | 0.1203 | 45 639 | 11 890 |
| 13. Tapi | 65 145 | 0.2284 | 0.1255 | 14 879 | 8 173 |
| 14. West flowing rivers from Tapi to Tadri | 44 940 | 1.5626 | 0.1694 | 874 110 | 9 479 |
| 15. West flowing rivers from Tadri to Kanya Kumari | 56 177 | 2.0210 | 0.1568 | 113 532 | 8 810 |
| 16. East flowing rivers between Mahanadi and Pennar | 866 643 | 0.2599 | 0.2630 | 22 520 | 22 788 |
| 17. East flowing rivers between Pennar and Kanya Kumari | 1 100 139 | 0.6440 | 0.2088 | 16 458 | 29 902 |
| 18. West flowing rivers of Kutch, Saurashtra including Luni | 321 851 | 0.0469 | 0.0433 | 15 098 | 13 948 |
| 19. Area of inland drainage in Rajasthan desert | | – | – | – | – |
| 20. Minor rivers draining into Bangladesh | 36 302 | 0.8539 | – | 31 000 | – |

**Table 2**

**Freshwater use for various purposes in India (million ha-meters)**

| Use/year | 1985 | 2000 | 2025 |
|---|---|---|---|
| Irrigation | 47.0 | 63.0 | 77.0 |
| Domestic and livestock | 1.7 | 3.3 | 4.6 |
| Industries | 1.0 | 3.0 | 12.6 |
| Thermal power | 0.3 | 0.3 | 0.4 |
| Miscellaneous | 4.0 | 5.4 | 11.0 |
| **Total** | **54.0** | **75.0** | **105.0** |

## 2.2 Groundwater resources

The total replenishable groundwater potential is considered to be 431.4 km³ per year. After making provisions for drinking and industrial water supply and for other purposes (other than irrigation), which is about 15 percent of the total potential, the potential available for irrigation is 360 km³ per year.

**Table 3**

Inland water resources (x$10^5$ ha) of various types (1993) by state

| State/U.T. | Length of rivers and canals (km) | Reservoirs | Tanks, lakes & ponds | Beels, lakes & derelict waterbodies | Brackish water | Total |
|---|---|---|---|---|---|---|
| Andhra Pradesh | 11 514 | 2.34 | 5.17 | – | 0.64 | 8.15 |
| Arunachal Pradesh | 2 000 | – | 0.01 | 0.03 | – | 0.04 |
| Assam | 1 820 | 0.55 | 0.21 | 1.10 | – | 1.86 |
| Bihar | 3 200 | 0.60 | 0.95 | 0.05 | – | 1.60 |
| Goa | 250 | 0.03 | 0.03 | – | – | 0.06 |
| Gujarat | 3 865 | 2.43 | 0.71 | – | 0.95 | 4.09 |
| Haryana | 5 000 | NEG | 0.10 | 0.10 | – | 0.20 |
| Himachal Pradesh | 3 000 | 0.40 | 0.01 | – | – | 0.41 |
| Jammu & Kashmir | 27.78 | 10.07 | 0.17 | 0.06 | – | 0.30 |
| Karnataka | 9 000 | 2.11 | 3.52 | – | 0.08 | 5.71 |
| Kerala | 3 092 | 0.30 | 0.03 | 2.43 | 2.43 | 5.19 |
| Madhya Pradesh | 20 661 | 2.91 | 1.17 | – | – | 4.08 |
| Maharashtra | 3 200 | 2.79 | 0.32 | – | 0.10 | 3.21 |
| Manipur | 3 360 | 0.01 | 0.05 | 0.40 | – | 0.46 |
| Meghalaya | 5 600 | 0.08 | 0.02 | NEG | – | 0.10 |
| Mizoram | 1 743 | – | 0.02 | – | – | 0.02 |
| Nagaland | 1 600 | 0.17 | 0.50 | NEG | – | 0.67 |
| Orissa | 4 500 | 2.56 | 0.64 | 1.80 | 4.33 | 9.53 |
| Punjab | 15 270 | NEG | 0.07 | – | – | 3.00 |
| Rajasthan | N.A. | 1.20 | 1.80 | – | – | 0.03 |
| Sikkim | 900 | – | – | 0.03 | – | 0.03 |
| Tamil Nadu | 7 420 | 0.53 | 2.24 | 5.24 | 0.56 | 8.57 |
| Tripura | 1 200 | 0.05 | 0.12 | – | – | 0.17 |
| Uttar Pradesh | 31 200 | 1.50 | 1.62 | 1.33 | – | 4.45 |
| West Bengal (p) | 2 526 | 0.17 | 0.30 | – | 1.15 | 1.46 |
| Chandigarh | 2 | – | NEG | NEG | – | – |
| Dadra & N.H. | 54 | 0.05 | – | – | – | 0.05 |
| Daman &Diu | 12 | – | – | – | – | 0.04 |
| Lakshadweep | – | – | – | – | – | – |
| Pondicherry | 247 | – | NEG | 0.01 | 0.01 | 0.02 |
| **Total** | **173 287** | **20.90** | **22.54** | **13.00** | **12.35** | **68.79** |

There has been a steady increase in the irrigation potential from groundwater the use of which has gone up from 6.5 mha in 1951 to 35.38 mha in 1992. The contribution of groundwater to total food grain production of the country is significant, as more than 50 percent of the irrigated area is using groundwater and in several districts it is more than 80 percent.

Overexploitation of groundwater resources has caused continuous decline in water level, decline of well yields, drying of shallow wells, deterioration of groundwater quality, seawater intrusion into coastal aquifers and increase in cost of energy required to lift water from a greater depth.

The underground aquifers are supplemented from sources other than rainfall, such as seepage from canals and field channels, ponds, tanks, influent drainage from rivers and deep percolation from irrigated fields. In most of the canal-irrigated areas of the country, a substantial component of the water applied to fields percolates below the root zone and contributes to the groundwater potential. Based on the total length of main canals and distributaries, the total possible recharge in India due to canal seepage in estimated to be 5.463 mha-metres per year.

Conjunctive use of the surface and groundwater is being practised in India to a limited extent. The practice has been prevalent in the Cauvery delta in Tamil Nadu, the Godavari Pravara Canal system in Maharashtra, the Ganga and the Western Jamuna Canal in Haryana and in parts of Punjab and Uttar Pradesh.

In Israel and USA, saline groundwater has been used for irrigation to a limited extent after diluting it with good quality surface water. There is scope for adopting such practices in certain areas of Gujarat and Rajasthan where the groundwater is brackish.

## 2.3 Present water use

By 1994 utilization of water (surface and groundwater) was about 606 billion m³, leaving 536 billion m³ as unutilized. Of this, the use for irrigation, domestic and industrial purposes and for hydropower production and other purposes was 501.0, 30.4, 20.0, 34.0 and 20.0 billion m³, respectively. Statewise net irrigated area according to sources is given in Table 4.

The pre-eminent role of water in the environment is well reflected in the India's National Water Policy, adopted in 1987, which emphasizes that the preservation of the quality of the environment and the ecological balance should be a primary consideration for the sustainable development of water resources. Since 1978, techno-economic appraisal of water resource projects in India includes an assessment of environmental impact. Due to a monsoonal climate, about 40 mha of the land in India is prone to floods.

## 2.4 Farming practices

Both agronomic (contour farming, mulching, tillage, strip cropping and mixed cropping) and mechanical measures (contour bunding, graded bunding, bench terracing on steep slopes) are deployed, depending upon soil and climate. A simple practice of farming operations, viz. ploughing, cultivation and planning across the slope at the same elevation, is called contour farming. The ridges and furrows and the rows of the plants placed across the slope form a series of innumerable miniature reservoirs and barriers to the water moving over the land surface, reducing run-off, soil erosion and loss of plant nutrients.

Maintenance of optimum soil fertility using fertilizers has been receiving great attention, which is evident from the increase in fertilizer inputs from nearly 0.07 million tonnes in 1950 to 14.4 million tonnes in 1996-97. This has greatly contributed towards increasing the crop production.

**Table 4**

Statewise net area irrigated (thousand hectares) according to sources, 1992-93 (provisional)

| State/UT | Canals | Tanks | Tube wells | Other wells | Other sources | Total |
|---|---|---|---|---|---|---|
| Andhra Pradesh | 1 727 | 729 | 384 | 1 027 | 162 | 4 029 |
| Arunachal Pradesh | – | – | – | – | 36 | 36 |
| Assam | 71 | – | – | – | 210 | 572 |
| Bihar | 934 | 128 | 1 621 | 80 | 581 | 3 344 |
| Goa | 8 | | 15 | – | – | 23 |
| Gujarat | 557 | 26 | 552 | 1 504 | 3 | 2 642 |
| Haryana | 1 359 | 1 | 1 238 | 1 | 29 | 2 628 |
| Himachal Pradesh | 7 | 1 | 6 | – | 83 | 99 |
| Jammu & Kashmir | 132 | 2 | 1 | 1 | 19 | 311 |
| Karnataka | 903 | 257 | 243 | 482 | 309 | 2 194 |
| Kerala | 103 | 48 | 66 | | 114 | 335 |
| Madhya Pradesh | 1 684 | 177 | 482 | 1 840 | 590 | 4 775 |
| Maharashtra | 530 | 385 | – | 1 348 | 175 | 2 470 |
| Manipur | | | – | – | 65 | 65 |
| Meghalaya | – | – | – | – | 45 | 45 |
| Mizoram | – | – | – | – | 8 | 8 |
| Nagaland | – | – | – | – | 60 | 60 |
| Orissa | 938 | 1 440 | 298 | 536 | – | 2 070 |
| Punjab | 1 365 | – | 2 387 | 11 | 98 | 3 861 |
| Rajasthan | 1 428 | 463 | 469 | 2 335 | 32 | 4 471 |
| Sikkim | – | – | – | – | 16 | 16 |
| Tamil Nadu | 851 | 231 | 174 | 1 027 | 17 | 2 698 |
| Tripura | 29 | 1 600 | 11 | – | 2 | 50 |
| Uttar Pradesh | 322 239 | 074 | 7 140 | 498 | 361 | 11 322 |
| West Bengal | 717 | 1 376 | 689 | 23 | 219 | 1 911 |
| Andaman & Nicobar | – | – | | – | | – |
| Chandigarh | – | – | 1 | – | | 2 |
| Dadar & Nagar haveli | 2 | – | – | 1 | 1 | 4 |
| Delhi | 3 | – | 33 | – | – | 36 |
| Daman & Diu | – | | – | – | 1 | 1 |
| Pondicherry | 9 | – | 14 | – | – | 23 |
| **All India** | **17 084** | **3 243** | **15 824** | **10 714** | **3 236** | **50 101** |

## 2.5 Water management for individual crops

### Rice

It is estimated that out of 42.8 mha area cultivated under rice in the country, about 6.8 mha are rainfed uplands, 21 mha are irrigated medium lands, and 15 mha are rainfed lowlands, where water depth varies from 5 cm to more than 100 cm under natural conditions. The storage of rain water in rice fields enhances its utilization in crop production, and induces groundwater recharge.

One of the objectives of storing rain water in rice fields is to minimize irrigation water application. In an experiment, irrigation was applied after one day of disappearance of ponded water. Under such a water regime, four irrigations were required to the rice crop grown in the plots having 6 cm bund height, three irrigations to 10 and 14 cm bund heights, and two irrigations to plots having 18, 22 and 30 cm bund heights.

### Wheat

The Indo-Gangetic plains, comprising the states of Uttar Pradesh, Punjab, Haryana, Rajasthan, Madhya Pradesh and Bihar, form the most important wheat-growing areas. During 47 years wheat production increased about 11-fold, from 6.39 million tonnes in 1949-50 to 69.27 million tonnes in 1996-97. During the same period the area increased from 9.8 mha to 25.93 mha and yield from 655 kg/ha to 2 671 kg/ha.

About 85.2 percent of the total wheat sown is irrigated (1994-95). Wheat is grown largely under irrigation in most of the states. Efficient management of water seems to hold the key, since the use of fertilizers and other inputs is largely linked with it.

### Maize

Maize is one of the important crops of India. The area under maize cultivation remained almost the same since 1968-69 (5.72 mha) to 1996-97 (6.25 mha). However, the production has increased by 85 percent, i.e. from 5.70 million tonnes in 1968-69 (yield 0.997 tonnes/ha) to 10.61 million tonnes (yield 1.698 tonnes/ha) in 1996-97. This has been achieved by extensive use of improved technologies, comprising high-yielding varieties, fertilizers, and irrigation. The irrigated area under maize has increased from about 0.41 mha in 1949-50 to 1.259 mha in 1994-95 (20.5 percent of total area).

### Pulses

Pulses form a major part of food and a source of protein for a large section of the vegetarian population of the country. Pulses are mostly grown on marginal lands under rainfed conditions and are considered as poor farmers crop. In 1994-95, only 12.7 percent of the total area under pulse production (23.0 mha) was irrigated.

### Pigeon pea (arhar)

The results of experiments involving irrigation at critical growth stages have shown that the main *kharif* crop responded to up to two irrigations (6-7 cm each) at pre-flowering/flowering and pod-development stages. However, under limited availability of irrigation water, one irrigation could be provided either at flowering stage, which enhances the yield by 0.40 tonnes/ha over the rainfed treatment (0.42 tonnes/ha) at Morena, or at pod-development stage for which a 14 percent increase in yield over rainfed treatment (1.42 tonnes/ha) was recorded at Faizabad.

**Gram**

Gram has been found to respond to irrigation at branching, flowering and pod-formation stages at Nafasari where it yielded 2.01 tonnes/ha, which is superior to two irrigations provided at any two stages (1.62 to 1.72 tonnes/ha). One irrigation applied at pod–formation stage yielded 9.5 percent more than the watering at branching (2.01 tonnes/ha) at Delhi, whereas at Kota the flowering stage was found more responsive to irrigation which yielded 19.3 percent more than watering at branching (2.02 tonnes/ha).

**Mung bean**

Mung bean is one of the important *kharif* season pulse crop of India. At Jammu two irrigations (6 cm each) applied at presowing and 20 days after sowing were found optimal for higher yield (1.56 tonnes/ha) during summer.

**Black gram (urd)**

Black gram is a highly prized pulse, rich in phosphoric acid. Its annual production in the country is about 1 million tonnes from about 2.5 mha area. A study conducted at Bhavanisagar in sandy loam soils has shown that the optimum value of IW/CPE was 0.6 for *kharif* season crop, involving 8 irrigations (4 cm each) besides a presowing irrigation (6 cm) to ensure proper germination, which yielded 1.08 tonnes/ha.

**Field peas**

The crop is grown in Uttar Pradesh, Punjab, Haryana, Madhya Pradesh, Jammu and Kashmir, Bihar, Delhi, Maharashtra, West Bengal, northeastern Himalaya states and Tamil Nadu. If winter rains are insufficient it needs one or two irrigations at appropriate growth stages. One irrigation at flowering gave optimum yield of 1.75 tonnes/ha at Pantnagar. Similarly, at Kota one irrigation at either flowering or pod-formation stage produced 2.39 tonnes/ha yield which was superior to one irrigation applied at branching that yielded 2.01 tonnes/ha.

**Lentil (masoor)**

Lentil is an important crop of the winter season and is grown throughout the northern and eastern regions. It is mainly grown on low-lying marginal lands that are subject to inundation. Lentil is normally cultivated as a rainfed crop, but it responds favourably to one or two irrigations depending upon rainfall and initial soil moisture storage in the profile.

**Oil seed crops**

Oil seeds, though energy-rich crops, are grown under energy starved conditions. The area under all the oil seed crops has increased by 2.5 fold between 1950-51 (10.7 mha) and 1996-97 (26.8 mha); at the same time the production has increased from 5.16 million tonnes to 25.0 million tonnes, which is nearly 5 fold. The increased productivity (0.48 tonnes/ha to

0.80 tonnes/ha) during the period is attributed to the use of improved varieties and management practices, including better use of irrigation water and fertilizers. Ground nut, rapeseeds and mustard, sesame, coconut, linseed, sunflower and soybean are the major oil seed crops grown in India.

**Sugar cane**

Sugar cane is the major source of sugar in the country. It is an important tropical crop, whose cultivated area has grown about 2.4 fold from 1.71 mha in 1950-51 to 4.17 mha in 1996-97. During the same period the area under irrigated cane has expanded from 67.3 percent to 87.9 percent, thus registering a 2.7-fold increase. Increased irrigated cane area in conjunction with other improved production technologies have steadily boosted the production from 57 million tonnes to 277 million tonnes.

**Cotton**

Cotton is one of the most important commercial fibre crops playing a key role in the economic and social affairs of the country. India ranks first in the world in respect of area under cotton and fourth in its production. The area under its production has expanded from 7 mha during 1953-54 to 9.12 mha during 1996-97. The production has increased from 4.13 million bales (of 170 kg each) to 14.25 million bales during the same period. The irrigated cotton area has also gone up from 10 percent in 1955-56 to 34.2 percent in 1994-95, which, together with other improved production technologies, has contributed to the increase in production.

Water requirements of crops vary with the crop and its variety, soil condition, fertilizer schedule and climate. Table 5 presents the rate of peak consumptive use expected under different climatic conditions.

**Table 5**

Maximum rates of soil moisture use by crops under different climatic conditions

| Climatic conditions | Peak rates of soil moisture removal m/day |
|---|---|
| Cool, humid | 3 |
| Cool, dry | 4 |
| Moderate, humid | 4 |
| Moderate dry | 5 |
| Hot humid | 5 |
| Hot dry | 8 |

## 2.6 Methods of irrigation

Efficient methods of irrigation enable the application of the right amount of water to the crop at the right time and its uniform distribution in the field. Irrigation water may be applied to the crops by spreading it in a sheet or in small streams on the soil surface, by spraying it over the crop through pipes and nozzles or by applying it in drops to wet the crop root zone.

**Border irrigation**

Border irrigation is a controlled surface flooding method of water application. The field is divided into a number of long parallel strips called borders, that are separated by low ridges. The width of a border usually varies from 3 to 15 m. The length of a border varies from 60 to 120 m in sandy and sandy loam soil, 100 to 180 m in medium loam soils and 150 to 300 m in clay loam or clay soils. The border method of irrigation is most suitable for close growing grain crops, such as wheat, barley and fodder crops. Field trials in India have shown that border irrigation is suitable in most areas of the country. Water is retained in the basin until it soaks into the soil.

**Basin irrigation**

The basin method of irrigation is essentially the check basin method applied to orchards. In flat lands, the basins are square, while on sloping lands the basins are formed by constructing ridges along adjacent contours with short cross ridges at either end. Usually there is one tree to a basin.

**Sprinkler irrigation**

In sprinkler irrigation, water is sprayed into the air through a sprinkler nozzle and allowed to fall on the land surface in a uniform pattern at a rate less than the infiltration rate of the soil. Sprinklers were introduced in India during the early 1950s. Initially, the sprinklers were used on high value plantation crops such as tea, coffee, chicory, cardamom and in orchards. Their use is gaining popularity on food crops, orchards, cotton and vegetables in areas where sprinklers are economically justifiable and technically feasible. Sprinkler irrigation can be used for almost all crops (except rice and jute) and on nearly all soils.

**Drip irrigation**

Drip irrigation, also called trickle irrigation, involves slow application of water to the plant root zone. The losses by deep percolation and evaporation are minimized. A precise amount of water is applied to replenish the depleted soil moisture at frequent intervals, for optimum plant growth. The system enables the application of water and fertilizer at an optimum rate to the plant root system. The amount of water supplied to the soil is almost equal to the daily consumptive use, thus maintaining a low moisture tension in soil.

Future gains in agricultural productivity of the country shall be critically determined by proper development and utilization of its surface and groundwater resources. Technological developments in pump manufacturing and deep well drilling have made groundwater extraction a common phenomenon, and the last three decades have witnessed a substantial increase in the groundwater development and utilization. Presently, groundwater is used to irrigate almost 50 percent of the total irrigated land in the country through about 5.5 million tubewells and more than 10 million dug wells. Unfortunately, these advantages have led to overexploitation and unsustainable utilization of this precious natural resource.

## 2.7 Management of poor quality water

Selection of crops, their varieties and salt tolerance at various stages of their growth are also important. Detailed guidelines for using saline irrigation water in India are available.

In the extreme cases of waterlogging and salinity, the seeds may not germinate and the plants may wilt permanently. Waterlogging and salinity of agricultural lands are due to a natural process or man-made activities. In irrigated areas, chemical degradation of land is a subsequent development on account of long duration waterlogging. Besides the erratic rainfall distribution, inundation of coastal areas by back-water flow from the sea and absence of proper soil and water-conservation measures in the catchments are also important factors leading to drainage congestion and related problems in the agricultural lands.

There is a lack of up-to-date and methodically monitored information on the type, severity and extent of drainage problems in the country. Estimates of waterlogged areas in India range from 2.5 to 16 mha, and those for salinized areas from 3.3 to 10.9 mha.

Beyond the above gross estimates for the whole country, very few published records based on rigorous monitoring exist on the progressive development of waterlogging and salinity in India.

## 2.8 Management of rain water resources

The technologies of rain water management are highly location-specific and are determined by physiographic, environmental, technical and socio-economic reasons. The main features are to conserve rain water where it falls, i.e. in the soil profile up to its maximum capacity and in storage structures for crop irrigation at a later stage. The technologies practised are (i) in situ water conservation, (ii) supplemental irrigation by run-off recycling, (iii) watershed management including agriculture, (iv) pond culture adjunct to canal system, (v) construction of high tidal-dykes, (vi) rice-fish integrated farming, (vii) using pressurized irrigation system, (viii) groundwater management, (ix) flood water management, (x) management of drainage and water-logging. Needless to say, ultimately the approaches have to be compatible with socio-economic adjustments of the target population.

# THE USE OF IRRIGATION SYSTEMS FOR FISH PRODUCTION IN KYRGYZSTAN

by

**D. Djancharov**
**Kiev Street 96a, Bishkek, Kyrgyzstan**

## ABSTRACT

In the mountainous country Kyrgyzstan the main source of water is ice- and snowmelt. There is a large natural lake Issyk-Kul, several smaller lakes, a number of reservoirs, of which Toktogul is the largest, and a number of rivers. The original fish stocks have been enhanced by introductions of exotic species, which gradually became dominant in fish catches. Regular stocking of Issyk-Kul (Sevan) trout (*Salmo ischchan issykogegarkuni*) has been practiced in lakes Issyk-Kul, Son-Kul and in Toktogul reservoir. Coregonids have been also introduced and have established self-propagating fish stocks. The introduced Chinese carps, common carp and pikeperch are several other fish of commercial importance which form an important component of fish catches from some waterbodies. From the native fish two species of dace (*Leuciscus*) are also important components of fish catches. Lake Issyk-Kul fish stocks are regularly enhanced by fish seed and fingerlings produced in two hatcheries/fish farms. The dissolution of the Soviet Union has led to a gradual decline in fisheries infrastructure, which formerly came from other states of the Soviet Union. Shortage of funds has resulted also in shortage of formulated feed which prevents fish production from being rehabilitated to the level prior to independence. The future of Kyrgyzstan fisheries appears to be in fish pond aquaculture.

## 1. INTRODUCTION

The Kyrgyz Republic or Kyrgyzstan is situated in the northeastern part of Central Asia. It borders Kazakhstan in the north-west, China in the south-east, Tajikistan in the south, and Uzbekistan in the west. Kyrgyzstan covers 198 500 km$^2$, and is a mountainous country, with a range of altitudes from 400 m to 7 400 m. About half of the country's territory is situated at an altitude of more than 3 000 m above sea level. The mountain masifs of Pamir-Alai and Tien Shan are the dominant features.

The climate of the country is strictly continental. The average January temperatures range from –1 °C to –8 °C in the valleys, and down to –27 °C in the mountains. Average July temperature ranges from +15 °C to +27 °C in the valleys and averages around +5 °C in the mountains. The average precipitation is from 180–250 mm/year in the eastern Tien Shan to 900–1 000 mm/year on the southwestern slopes of the Fergana mountains.

Small and large glaciers cover more than 8 000 km$^2$. They contain about 700 km$^3$ of water, which represents the main water source for rivers and groundwater. The volume of groundwater resources is estimated at 13.7 km$^3$. The volume of the prospected and approved resources of 44 fields of fresh groundwater that can be utilized annually is 4.85 km$^3$.

There are more than 3 500 rivers in the country, which belong to seven main river basins: Syr-Darya, Amu-Darya, Chu, Talas, Kurkure-Suu, Tarim, Ili, and to Lake Issyk-Kul (Isek-Kol). There is only about 20 percent difference in annual mean river discharge, due mainly to variations in melting of glaciers and snow.

Most rivers have a low winter discharge. On average the water discharge during the vegetation period represents 74 percent of the total annual discharge, compared to only 26 percent during the rest of the year.

Rivers receiving ice- and snowmelt water have approximately 60–85 percent of the annual discharge from March to August, with a peak in May. This type is typical for the rivers of the Fergana valley, e.g. the rivers Donguztau, Zerger, Yassy, Ugart. Water from these rivers is used for irrigation from the end of April or beginning of May, and water consumption reaches its peak in July/August. This type of water use is typical for the high and central part of the Naryn Basin and the basins of the Kyzyl-Suu and Tarim rivers.

The use of water from rivers with a mixed type of water supply starts increasing in March/April and reaches its peak in July/August, decreasing in September. This type is typical for example for the basins of the Chu and Talas rivers and the lower part of the Naryn basin.

Annual discharges of rivers originating in Kyrgyzstan are given in Table 1.

**Table 1**
Average annual discharges from water basins of Kyrgyzstan

| Basin | Average annual discharge ($km^3$) | Average annual discharge on the territory of Kyrgyzstan ($km^3$) |
|---|---|---|
| Syr-Darya | 46.04 | 27.40 |
| Amu-Darya | 93.42 | 1.93 |
| Chu | 6.64 | 5.00 |
| Talas and Kurkure-Suu | 1.84 | 1.74 |
| Ili-Kar-Kara | 0.36 | 0.36 |
| Issyk-Kul lake | 4.65 | 4.65 |
| Tarim | 6.15 | 6.15 |
| TOTAL | 159.10 | 47.23 |

The table shows that most water forms in the basins of :
– Amu-Darya and Syr-Darya (93.42 – 46.04 $km^3$), and in
– Kyrgyzstan in the Syr-Darya, Tarim and Chu basins (27.4 – 61.15 – 5.0 $km^3$).

Prior to the 1930s Kyrgyzstan waterbodies contained fish of seven families and 60 species and subspecies, the important species being sazan (wild form of *Cyprinus carpio*), osman (*Dyptichus dybowskii*), marinka-snowtrout (*Schizothorax pseudaksaiensis issykkuli*) and chebak (dace) (*Leuciscus schmidti*) in Lake Issyk-Kul, and pike (*Esox lucius*), catfish (*Silurus glanis*), marinka (*Schizothorax issykkuli tschuensis*), sazan and asp (*Aspius aspius*) in rivers and ponds of the Chu valley.

To increase fish production in waterbodies of Kyrgyzstan as well as to bring in more economically valuable fish species compared to some of the indigenous fish, introductions were carried out starting in the 1930s. Sevan trout (*Salmo ischchan gegarkuni*) and whitefish (*Coregonus lavaretus*) were transferred from Lake Sevan in Armenia to Lake Issyk-Kul. Omul (*Coregonus autumnalis*), peled (*Coregonus peled*) and least cisco (*Coregonus sardinella*) were brought from Siberia, and other fish from elsewhere included khramulya (*Varicorhinus capoeta*), carp, pikeperch (*Stizostedion lucioperca*), whitefish (*Coregonus* spp.) and bream (*Abramis brama*). As a result of breeding Sevan trout in the Ton fish hatchery for stocking in Lake Issyk-Kul, their numbers there increased, and starting in 1975 fishing for this species was permitted in the lake.

According to the Kyrgyzstan President's decree of 25 July 2000 and the resolution of Kyrgyz Government "Improving the government structure of the Kyrgyz Republic", the implementation of government policy in the spheres of development and coordination of water resources, organization of fishery, fish production and fish protection, was delegated to the Ministry of Agriculture, Water and Processing Industry of the Kyrgyz Republic. A new unit was created in the ministry, called Sector of Fishery Industry. Its major objective is to improve fisheries management.

At present there are two complete fish farms, i.e. Uzgen and Talas, and Ton hatchery. The farms do not work at their full capacity because of lack of funds. The Uzgen fish farm started in 1968, with fish ponds constructed to produce 500 tonnes of fish annually. The 91 ponds cover 290 ha. The Talas fish farm started in 1975. It has 364 ha of ponds and an annual fish production capacity of 600 tonnes.

Formerly the state fisheries was the major producer of fish in Kyrgyzstan: Uzgen fish farm produced 400–500 tonnes and Talas fish farm 150 tonnes. The break of economic links with the former Soviet republics has led to shortage of hatchery equipment and special fish feeds. The government-run fisheries is at present in a critical situation, also due to shortage of funds, and this has been reflected in the decline in fish production. Ponds with low production capacity have been converted for cultivation of agricultural crops, such as tobacco, rice and maize to raise money for fisheries. But this is not sufficient. In order to achieve the previous level of fish production, the fisheries needs foreign investment.

At present the Talas and Uzgen state fish farms are producing carp, sazan and grass carp. The Talas fish farm produces approximately 45 000 two-year-old carps, the Uzgen farm about 200 000 carp fingerlings and 71 000 older fish. Private farmers grow fish in small waterbodies of Chu Oblast at Kara-Tuma, Puchuk, Maltabar, Stepninsky and Agermen, but further development of fish pond culture depends on improving the supply of fish feed. One of the main reasons for the dramatic decrease in fish pond production is the lack of suitable fish feed.

Recently, for the first time in two years, induced breeding of fish has been restarted in lakes Issyk-Kul and Son–Kul. Ton hatchery produces Issyk-Kul (Sevan) trout, for which it has an incubation capacity of about 10 million fry, as well as being able to produce 1 million of carp fry. However, on-growing is limited due to the shortage of suitable feed. In 1998, 398 000 young trout of 140 mg each were released, and in spring of 1999, 285 000 of 160 mg each were released. The planned quantities of whitefish were also produced. On 13 April 2000 there were further releases of Issyk-Kul trout into the lake, some of them from the Karakol fish farm. But to make both fish farms more effective and working at full capacity long-term favorable credits and grants are needed.

In Lake Son-Kul 18 200 whitefish eggs were fertilized and later on released to hatch naturally in the lake. To stop illegal fishing, in the year 2000 controls of fish on sale and of records on fish marketing were introduced. Fishing units have also received a contract for fishing specific areas. Fishery inspections now control the records on fish catches and fish production on fish farms. Fish farms and fishing units are now given technical and advisory support.

Kyrgyzstan's fisheries badly needs financial support which has not been forthcoming for a number of years. Assistance is needed for the purchase of formulated feed, for rehabilitation of infrastructure of hatcheries and fish farms, for equipment as well as for fish transport.

At present commercial fishing is carried out mainly on lakes Issyk-Kul, Son–Kul, and on Toktogul reservoir. Uzgen and Talas fish farms produce common carp, grass carp and silver carp. During the period 1991–2000 the total fish catch from these three large waterbodies was 3 527 tonnes.

## 2. LAKE ISSYK-KUL

Lake Issyk-Kul is the largest lake in Kyrgyzstan and a major fisheries water body. Each year 1 000 to 1 200 tonnes of high quality fish are captured here. But according to the Issyk-Kul biological station of the Academy of Sciences, commercial fishery should not harvest more than 250–300 tonnes.

Lake Issyk-Kul is situated at an altitude of 1 608 metres, the lake surface area is 6 206 km$^2$, average depth 279 m, and maximum depth 668 m. Shallows make up one fourth of the reservoir surface area. The lake is 179 km long and 60 km wide. It is saline, and with the exception of the bays Tyup and Rybachi, it is not covered with ice in winter. The maximum surface water temperature in summer reaches 21 °C.

Issyk-Kul is an internal drainage waterbody, with 59 large and small rivers entering it, but without an outflow. Most of the inflows are used for irrigation of agricultural lands in summer. The largest inflowing rivers are Karakol, Jergalan, Tyup and Barskaun.

The lake has 23 fish species, of which 12 are indigenous, 11 intentionally or incidentally introduced. The dominant commercial fish species are chebachok (*Leuciscus bergi*), pikeperch (*Stizostedion lucioperca*), chebak (*Leuciscus schmidti*), whitefish (*Coregonus* spp) and Issyk-Kul trout (*Salmo ischchan issykogegarkuni*). At present their stocks are satisfactory. There are good catches of both species of *Leuciscus*, but lower catches of the whitefish *Coregonus lavaretus*.

There is a potential for increasing catches of both species of *Leuciscus* as well as *Coregonus lavaretus* through exploitation of lake areas which are not used now. *L. bergi* is the main species fished and it dominates the catches. The introduced pikeperch is well distributed in all freshwater parts of the lake and puts a heavy pressure on both species of *Leuciscus* on which it feeds. There is therefore no limit placed on fishing for pikeperch which can be fished all year around. The main stocks of *L. schmidti* are in the western and eastern parts of the lake.

Fishing of the introduced whitefish started in 1982, and since the end of the 1980s its annual catches have ranged from 20 to 30 tonnes. It has established self-propagating stocks. The stocks of the Issyk-Kul (Sevan) trout are maintained by regular stocking, but the lake also has small self-propagating populations.

## 3. LAKE SON-KUL

The high altitude Lake Son-Kul is the largest freshwater natural waterbody in Kyrgyzstan. The lake is situated in the central Tien Shan at an altitude of 3 016 m. It covers an area of 273 km$^2$, and has a mean depth of 9.3 m.

Until 1959 the lake contained only aquatic invertebrates. Later on several fish species were introduced. These were hybrids of common carp and the wild form of the common carp, tench, Issyk-Kul trout, osmans *Dyptichus dybowskii* and *D. maculatus*, and Tibet loach *Nemacheilus stoliczkai*.

Fish introductions started in 1969 with stocking *Coregonus peled*. In 1974 *C. lavaretus* and *C. nasus* were added. At present the fish fauna of the lake is represented by the following species: *Coregonus peled*, *C. lavaretus*, *D. maculatus*, *N. stoliczkai*, and an occasional appearance of *Coregonus nasus*. The lake provides favourable conditions for *C. peled* and its fishing started in 1976. By that time the fish reached an average weight of 1 kg. Fish were captured in nets with a mesh size of 70 mm, and in 1976, 108 tonnes were harvested. The catches declined in the following years, with 45.4 tonnes harvested in 1990, 31.5 in 1995, and 24.6 in 2000. The decline in catches could have been caused by fishing during the breeding period, which hit the females before they had time to spawn.

## 4. TOKTOGUL RESERVOIR

The largest man-made water reservoir in Kyrgyzstan was constructed on the Naryn River as a storage for hydroelectricity production. The reservoir is at 900 m altitude and situated in the Ketmen-Tubin valley of Osh Oblast. Apart from the Naryn the reservoir also receives the rivers Chychkan, Uzun-Ahmat, Nichke and Kandul. It covers 284 km$^2$ and the full storage capacity is 19.5 km$^3$. The mean depth is 65 m, maximum depth 210 m. The reservoir is 70 km long and the maximum width is 13 km. The gorge in the western part of the reservoir is 6 km long and 90–95 m deep. The lake-like sector is much larger but also much shallower, with a maximum depth of only 15 m.

The average maximum water surface temperature is 20–23 °C, in winter 10 °C, pH is 7.5–8.0, concentrations of dissolved oxygen range from 9 to 13 mg/litre. The reservoir has a medium level of productivity.

The fish of the reservoir are derived from the fish stocks of the Naryn and its tributaries plus several introduced species. The local fish include *Schizothorax intermedius*, Syr-Darya dace (*Leuciscus squaliusculus*), asp (*Aspius aspius*), osman (*Dyptichus* spp.), stone loach

(*Nemacheilus barbatulus*) and Turkestan catfish (*Glyptosternum reticulatum*). The introduced fish are Amu-Darya trout (*Salmo trutta oxianus*), grass carp, silver carp, bighead carp (*Aristichthys nobilis*), sazan, common carp, Siberian sturgeon (*Acipenser baeri*) and Amur goby (*Rhinogobius similis*).

Commercial fishery started in 1978, but there were large differences between annual catches. During the first years (1978–1980) catches increased from 12.7 tonnes to 70 tonnes. A certain stabilization can be noticed since 1988, with catches reaching up to 25 tonnes. While initially the catches were dominated by *Schizothorax intermedius*, since 1989 the Amu-Darya trout has been the dominant species. In 1977, Issyk-Kul (Sevan) trout was introduced and, by 2000, 17.5 tonnes were captured, representing 68.9 percent of the total catch. The biology of this fish in Toktogul reservoir is not well known. According to the data collected during 1980–1990 its growth rate is similar to that in Lake Sevan. It reaches maturity at the age of 3–4 years, when 45–50 cm long. Females mature one year later than males. The number of eggs per female is 3 300 to 17 000 (average 6 500). The spawning starts usually in the second half of October at water temperature of 5.7–6.2 °C. It is probable that the trout has established a self-propagating population in this reservoir. It is believed that with regular stocking the sustainable annual catch could reach 120 to 150 tonnes.

The most suitable place for Issyk-Kul trout spawning is the river Uzun-Ahmat. But the trout often faces the problem of reaching spawning grounds. This needs to be resolved, in addition to enhancing hatchery production of a sufficient number of stocking material.

Snowtrout *Schizothorax intermedius* is a representative of the fish fauna of the Central Tien Shan. The new reservoir had favourable conditions for this species which dominated the catches during the early years. But a heavy fishing pressure on spawning stocks caused its decline to today's level of 2–3 tonnes per year.

Starting in 1978 herbivorous Chinese carps (grass and silver) have been stocked in order to utilise aquatic plants and phytoplankton. They have adapted well to the reservoir conditions and have a fast growth rate. They are now common, but they do not reproduce in the reservoir. In 1991, 2.5 tonnes of silver carp and one tonne of grass carp were captured. The sturgeon *Acipenser baeri* was introduced in the reservoir in 1982. Very little is known what happened afterwards. Today it is extremely rare.

## 5. OTHER RESERVOIRS

The following is a brief account of fish species found in less productive reservoirs of Kyrgyzstan:

Orto-Tokoy reservoir – snowtrout, osman, Issyk-Kul trout
Kirov reservoir – snowtrout, common carp, osman, grass carp, silver carp
Naiman reservoir – wild form of common carp, common carp
Bazar-Korgon reservoir – wild form of common carp, common carp.

The above listed irrigation reservoirs are not very suitable for commercial fish production because of the irregularities in water level and because of their low productivity. But they are suitable for recreational fisheries.

## 6. DISCUSSION

For the perspectives of fisheries development in Kyrgyzstan (see Table 2).

**Table 2**

Perspectives of fisheries development in Kyrgyzstan

| Waterbody, fish | 2001 | 2005 |
|---|---|---|
| Ponds, herbivorous fish | 50 tonnes | 300 tonnes |
| **Lakes** | | |
| – Issyk-Kul (*C. lavaretus*, pikeperch, *Leuciscus bergi*, *L. schmidti*, Issyk-Kul trout , bream) | 85 tonnes | 150 tonnes |
| – Son-Kul (*C. peled, C. lavaretus*) | 40 tonnes | 80 tonnes |
| – Kara-Suu (*C. peled, Schizothorax*) | 5 tonnes | 15 tonnes |
| **Reservoirs** | | |
| –Toktogul (herbivorous: carp, sazan, *Schizothorax*) | 40 tonnes | 150 tonnes |
| –Bazar-Korgon (carp, grass and silver carps, *Schizothorax*) | 10 tonnes | 30 tonnes |
| – Orto-Tokoy (*Schizothorax, Salmo trutta oxianus*) | 5 tonnes | 20 tonnes |
| **Total**: | 235 tonnes | 745 tonnes |

To improve the situation in fisheries the following assistance is needed:

- financial support for rehabilitation of fishery, especially fish hatcheries and fish farms,
- supply of equipment for hatcheries, fish farms and research,
- training in new methodologies,
- better control of fish farms,
- use of the land around ponds for cultivation of corn and other crops for fish feed to reduce the cost of fish production.

Production of marketable fish can also be increased by expansion of the pond areas, intensification of fish production, reconstruction of hatcheries and fish farms, provision of credit and perfection of aquaculture technologies. Qualified staff should be employed for hatcheries and fish farms and they should be trained in efficient aquaculture methods.

# THE IMPACT OF DROUGHT ON AGRICULTURE AND FISHERIES IN IRAN

by

M. Foghi
Ordibehest Avenue, Janb Omor Senfe, Esfahan – Shilat, Iran

## ABSTRACT

In the first part this paper discusses the impact of drought on water resources and agriculture in Iran. It briefly describes the various types of irrigation systems in use in Iran, and the effort to economise agricultural water use. The second part deals with capture and aquaculture fisheries, including the consequences of the recent prolonged drought on fish farming. The contribution of aquaculture to food security in the country is also discussed.

## 1. WATER CRISIS IN IRAN AND AGRICULTURE PRODUCTION

Iran is one of the countries with the earliest agriculture, with agriculture records going back to 4000 BC. But overuse of the soil has led to soil erosion and desiccation of many water sources. Iran gradually changed into a dry and drought-prone country and today it belongs to the arid/semi-arid zone of Asia.

The average annual rainfall in Iran is 243 mm as compared with the global annual rainfall average of about 835 mm. There are some areas in Iran which have less than 50 mm of rainfall. The rainfall is unevenly distributed throughout the year which prevents optimal exploitation of water resources, with the rain falling largely during the non-vegetative season. While one percent of the world population lives in Iran, Iran has just 0.36 percent of the total renewable freshwater resources of the world. Iran is already using 66 percent of its freshwater resources.

Of the estimated annual 400 billion $m^3$ water received as precipitation 260 billion $m^3$ are lost through evaporation and infiltration. This leaves about 130–140 billion $m^3$ to be used, out of which only 97 billion $m^3$ are exploited: 9 billion $m^3$ for drinking and other human uses, including industries, and the remaining 88 billion $m^3$ for agriculture. In 1961, the per capita water availability in Iran was about 6 500 $m^3$, which has decreased to 2 100 $m^3$ today, and the decline continues. It is estimated that by 2021 only 1 200 $m^3$ per capita will be available.

More than 50 percent of the country's freshwater resources are in the form of groundwater, which must be conserved for drought years. At present overexploitation of the water resources has resulted in some changes in water quality. The recent two-year drought (1998–2000) has led to overexploitation of the groundwater in all areas, and consequently in some areas salt water has entered the aquifers. There has also been an increase in water pollution. Soil disturbances such as those due to deforestation in the river catchments, water overexploitation and poor water quality are some of the major factors which intensify the scale of the crisis caused by drought.

Irrigated lands cover 8 million hectares (ha) and those under dry farming cover 6.5 million ha, but the latter category has been subject to serious damage imposed by the recent droughts: when comparing dry farming production areas of 1992–1993 with those in 1998–1999 there was a drop from 7.11 million ha to 4.46 million ha. As a result of that, during the same period the total area of farms and orchards declined from 14.36 million ha to 12.32 million ha. At the same time, the total agricultural production increased from 54.22 million tonnes in 1992–1993 to 60.2 million tonnes in 1998–1999.

The area under dry wheat farming was reduced by 45 percent, wheat production by 65 percent, and wheat yields by 37 percent when comparing the year 1993 with 1998–99. This has been the result of the decline in rainfall. To make up for the shortfall, Iran had to import 7 million tonnes of wheat in 1999–2000 which has made it one of the biggest wheat importers in the world.

Water shortage is considered a major factor responsible for the slow growth of agricultural production. Rational and scientific planning in order to secure maximum exploitation of water resources is strongly needed.

## 2. IRRIGATION SYSTEMS IN IRAN

The limited water resources and the widespread use of unscientific and traditional irrigation systems which result in up to three times higher water consumption than actually needed for production of crops, are the major constraints facing agriculture in Iran. To overcome this a new irrigation improvement programme is being implemented. Its objectives include increasing production rate in the irrigated farming system through rationalization of water resources. The programme focuses on reduction of the negative impacts of drought, using modern irrigation methods.

There are three common irrigation systems in Iran: surface irrigation, sprinkler irrigation, and gravity drip irrigation.

Surface irrigation system is the oldest and most common system applied. It does not require special equipment and machinery, and if well implemented it is considered to be one of the best systems in Iran. However the water use is wasteful and if poorly used crop yields decline. Surface irrigation is applied as basin irrigation, border irrigation or furrow irrigation.

Soundly managed traditional irrigation system requires 1–1.5 litre of water per second per hectare, while sprinkler systems use only 0.6–0.8 per second. This means that the same amount of water as used in traditional irrigation results in higher production. Sprinkling irrigates the area near the roots meaning that only a small area and only the required soil depth are wetted. This results in lower evaporation rate, as well as reducing the area suitable for weed growth.

Gravity drip irrigation has proved very satisfactory in irrigating plants in the desert, specifically in sandy soils where it is not feasible to apply common surface irrigation or sprinkling. It leads to accumulation of salt in the root area, which can be removed through continuous washing. As the plant receives fertilizers directly, then the efficiency of fertilizer use is increased as well as the efficiency of water use. This system produces good results with crops such as apples, pistachios, peanuts, grapes and watermelons. With sound management this method requires 0.4–0.6 litre/sec/ha. The method is more nature-friendly, but also more costly to apply.

To overcome the crisis resulting from the shortage of water available for irrigated agriculture, the following proposals have been formulated. If implemented, they should result in the urgently needed improvements and in higher crop production.

- Provision of training to farmers to raise their awareness of the shortage of water.
- Restoration of vegetative cover in catchments.
- Upgrading the performance of irrigation water.
- Prevention of the input of pollutants into water.
- Prevention of fragmentation of agriculture lands and encouraging farmers to integrate farming so as to prevent water wasting.
- Construction of dams to create water storage.
- Promotion of the use of saline waters in agriculture.
- Application of modern irrigation systems.
- Recycling sewage.
- Estimating exact water requirements of plants.
- Introducing plant species requiring less water.

## 3. AQUACULTURE IN IRAN

Having the advantage of about 2 700 km of coastline in the Persian Gulf, Oman Sea and the Caspian Sea, Iran has traditionally pursued capture fisheries in those waters. The production achieved from both the southern and northern waters has increased fifteen fold during the period of 1973–1998. Aquaculture started in 1922 with the breeding of sturgeon to produce fry for stocking in the Caspian Sea. This was followed by trout farming in 1959. Today the major aquaculture activities in Iran are well diversified.

<u>Warm water fish farming</u>

The following fish species are farmed: common carp, grass carp, silver carp and bighead carp that are mostly brought from abroad. Production of these species on fish farms has increased by 46 percent since 1989.

<u>Cold water fish farming</u>

Trout farming started in 1959 and the production increased from 599 tonnes in 1978 to 26 700 tonnes in 2000. Rainbow trout (*Oncorhynchus mykiss*) is the main cold water fish species farmed in Iran.

<u>Fish farming in natural and semi-natural water resources</u>

The total area of inland water bodies including lagoons, lakes, reservoirs, rivers, etc., in Iran is estimated to be about 1.5 million ha. Some of the water bodies are stocked with common, bighead and silver carps. Freshwater and Caspian Sea capture fisheries (including waters enhanced by stocking) have leveled off at around 34 000 tonnes/year, with a steep drop in 2000 probably due to the continuing drought conditions.

Fish stock enhancement

Fish fry and fingerlings are produced under aquaculture conditions for stocking in inland water bodies and the Caspian Sea. This is to compensate for the losses resulting from fishing and to boost the existing fish stocks. The following species are produced and stocked: kutum (*Rutilus frisii*), sturgeon, bream, trout, perch, *colmeh* and several species of carp. Over the period 1978–1999 the number of fish fry and fingerlings annually released into the Caspian Sea and inland waters rose approximately 15 fold .

Shrimp culture

Iran enjoys the advantage of a 2 000 km coastline in the south of the country. An area of 16 000 ha has been identified there as suitable for shrimp culture. Shrimp culture operations in Iran started several years ago and production has been increasing, reaching 4 000 tonnes in year 2000. *Penaeus indicus*, a native species of Iran, is the main cultured species and it yields up to 2 tonnes/ha.

Aquatic invertebrate culture

The focus is on the giant freshwater shrimp (*Macrobrachium rosenbergii*) and the slender clawed crayfish (*Astacus leptodactylus*) culture. The crayfish is produced for regular releases into natural water bodies. The giant freshwater shrimp is cultured in monoculture or with common carp.

# 4. CONSEQUENCES OF DROUGHT ON FISHERIES IN IRAN

The low rainfall in recent years has resulted in reduced fish production. When comparing the fish production from natural and semi-natural waters for 2000 with that in 1998, it shows a reduction by 34 percent. Warm water fish farm production was 20 percent lower in 1999 than in 1998, but in 2000 the production was close to the average for years 1995–1998. Cold water fish production has been also affected by the drought through reduced water flow rates in streams and rivers providing water to fish farms. While the total production from cold water fish farms was higher in 1999 than that in 1998, this was the result of opening 82 new fish farms. The average cold water fish production per fish farm during the drought year 1999 was considerably lower than that in 1998.

Drought has also had a negative impact on the shrimp aquaculture industry as a lower freshwater input raises water salinity in ponds and this has a negative effect on shrimp larvae and the final product. The total damage inflicted on prawn production in 2000 is estimated to be 33 000 million rials, of which at least 11 percent is believed to be the result of the drought impact. Lower river discharges into the sea reduce the spawning of shrimps which in turn results in lower shrimp catches. Over the last decade shrimp catches have decreased from 3 759 tonnes to 2 350 tonnes.

# 5. DIVERSIFICATION OF FISHERIES PRODUCTION GIVEN THE WATER LIMITATION

Iran like many other countries of the dry zone of Asia faces water shortages. The existing fisheries and its future development faces the problem of how to develop under water use restrictions. Several possibilities are discussed.

Fish farming using micro-water resources

A large volume of fresh water estimated at 53.3 billion $m^3$ is supplied from micro-water resources such as wells, springs and underwater aqueducts (*karez*). In many of these waters fish are present and fish production could be developed alongside the water used for agriculture and other purposes. Water stored in ponds fed from wells or other sources could support fish stocks. This would also facilitate promotion of aquaculture among the rural population, increase the per capita fish consumption among the farmers, generate income, create job opportunities, and by directing the organically enriched outflows from such fish farms onto fields, increase agricultural production. Production of warm water fish from such ponds reached 2 580 tonnes over the period 1993–1999.

Following the relative success with trout farming on fish farms, a number of coldwater fish production raceway ponds have been constructed near the micro-agricultural water resources. This has provided a good income to private farmers. The outflowing organically enriched water is used for irrigation agriculture. In 1998 coldwater fish farms established on this principle produced 800 tonnes of rainbow trout.

The use of micro-agricultural water resources for fish farming has been most welcome by farmers. The overall warm water fish production plan for 2021 is 454 000 tonnes, and for cold waters 180 000 tonnes. Of these 19 000 and 44 000 tonnes, respectively, should be produced using microwater resources.

Fish pond farms

One of the mechanisms for increasing fisheries production is the establishment of fish pond farms. These may be assisted by the government or be the result of private initiative. On warm water fish farms in Iran fish production is about 3 tonnes/ha, but on some farms it reaches up to 6 tonnes. Cold water fish farms currently in existence in Iran are mostly lacking an electricity supply which does not make possible water recirculation.

Stocking inland waters

There are about 3 500 inland water bodies. Of these about 730, covering a total area of 508 000 ha, are used for fisheries. During 1995–1999 the annual fish production from these water bodies averaged 30 000 tonnes, employing over 130 000 fishermen. It has been recognized that to keep the fisheries sustainable, regular stocking is necessary. Iranian fisheries together with the Ministry of Power is, subject to the Fishery Law, responsible for the coordination of the management of these resources, including stocking. Other enhancement methods have also been applied since 1970, such as cage culture. Fish are raised with some success in canals, with 129 tonnes and 108 tonnes of fish produced in 1998 and 1999, respectively.

Fish farming in rice fields

About 600 000 ha of land in Iran is allocated to rice farming. It is planned to further develop integrated rice-carp farming and trout farming during the post-harvest period. Trout farming would follow when the water temperature does not exceed 20 °C. Farming in rice fields in 1999 resulted in harvest of 126 tonnes of fish. The side benefits of this type of culture are fertilization of fields by fish and control of the rice stem borer by carp feeding on it.

Fish farming in brackish waters

Some areas in the Esfahan, Yazd, Qom, Khorasm, and Kerman provinces have water resources (mainly underground) of a high salinity, with electrical conductivity sometimes exceeding 30 000 micromhos. These waters which are unsuitable for agriculture could be utilised for fish farming. Since 1995 saline water of 10 000–25 000 micromhos has been used for rainbow trout farming in Iran. Trout production in saline water reaches about 3 tonnes/ha but with aeration it could reach 5–6 tonnes/ha.

*Macrobrachium rosenbergii* has been cultured in water of 20 000 micromhos, resulting in a production of 5–6 tonnes/ha.

## 6. IMPACT OF AQUACULTURE

This can be summarized as follows:

- food production and contribution to national food security
- reducing reliance on red meat and pressure on pastures
- contribution to sustainable economic development
- creation of job opportunities especially in rural areas.

**Number of employees in aquaculture sector (inland water farmers and fishermen)**

| Year | 89 | 90 | 91 | 92 | 93 | 94 | 95 | 96 | 97 | 98 |
|---|---|---|---|---|---|---|---|---|---|---|
| No. of employees | 9 957 | 9 923 | 10 934 | 10 204 | 10 532 | 10 921 | 11 004 | 11 630 | 10 250 | 16 661 |

- creation of regional development centers through establishing production farms on land unsuitable for agriculture
- development of aquaculture-related industries such as fish feed production, hatchery equipment manufacturing, processing and packaging industries
- substitution of farmed fish for captured fish to relieve the pressure on natural fish stocks
- increased per capita fish consumption
- strengthening the government economy through export of fisheries products.

There is a need to make fish accessible especially to the rural population.

**Annual per caput fish consumption (kg/person)**

| Year | 79 | 87 | 89 | 90 | 91 | 92 | 93 | 94 | 95 | 96 | 97 | 98 |
|---|---|---|---|---|---|---|---|---|---|---|---|---|
| Consumption | 1.0 | 2.5 | 4.25 | 4.4 | 4.4 | 4.25 | 4.5 | 4.0 | 4.5 | 4.5 | 4.5 | 4.6 |

Since 1989, the per caput fish consumption in Iran has stagnated at a level of around 4.5 kg/person/year.

# IRRIGATION AND FISH PRODUCTION IN MONGOLIA

**by**

**Dashdorj Ganbaatar**
**Room 50, Government House No 9, Enkhtaivan Avenue 16A,**
**Ulaanbaatar 210349, Mongolia**

## ABSTRACT

Because of the low precipitation (280–320 mm in the main irrigation areas, 90 mm in the Gobi and Steppe zones), rainfed crop production is limited. A simple irrigation system for 17 000 ha is now being used for crop production and the rest is used for pasture and hay production. At present Mongolia has no fish production in irrigation systems. Mongolia has 75 species of fish which belong to 36 genera and 11 families. There are over 30 species of commercial value. As the fishery resources in lakes and rivers are limited and easily overexploited there is a need for a rational management plan which would develop fishery as a sustainable food resource. More attention has to be paid to aquaculture, as well as to better preservation and processing of fish harvested by capture fisheries. At present Mongolia produces no fish in irrigation systems.

## 1. INTRODUCTION

Mongolia covers 1.5 million km$^2$ and is the fifth largest country in Asia. It is located in the north-central Asia between Russia and China. With an average altitude of 1580 m above sea level, it is one of the highest countries in the world.

Geographically, Mongolia can be divided into four regions, namely, Khangai forest region, eastern steppe region, Gobi desert region and the semi-desert region. Only 10 percent of the country is forested. The primary sources of water are the Altai, Khangai and Khentii mountain ranges. More than half of the rivers flow into the Arctic Ocean, the remaining flow into the enclosed drainage basin of Central Asia.

Mongolia has a continental climate which is characterized by extremes of heat and cold, with an average temperature of + 25 °C during summer and –25 °C in winter. The country is renowned for its blue skies with an average of 250 days of sunshine per year.

The country is rich in plants and wild animals, with an estimated 130 species of mammals, 370 species of birds and 75 species of fish. There are some 2 000 species of plants in Mongolia.

Mongolia has a relatively sparse population of some 2.6 million, which grows at annual rate of 2.5 percent. The population density is 1.5 person per km$^2$. Among the different ethnic groups Khalkha accounts for 72 percent of the total population. The official language is Mongolian which is spoken throughout the country. Buddhism is the official and most common religion.

The major export from Mongolia are livestock, meat, hides, skins and cashmere. A medium amount of gold and a large amount of copper are also exported. High transportation costs hamper export of other minerals.

Mongolia has more than 4 000 lakes of which only 200 have a surface area of more than 5 $km^2$. However, the few large lakes account for almost 90 perceny of the total water surface of Mongolia. Most Mongolian lakes are shallow, with depths rarely reaching 20 m.

The largest lake in Mongolia is Lake Khuvsgul in the far north of the country. It has a surface area of 2 760 $km^2$, volume of 381 $km^3$, and maximum depth of 262 m. It holds almost four fifths of all lake water in Mongolia.

## 2. IRRIGATION SYSTEMS AND AGRICULTURE IN MONGOLIA

Traditionally in Mongolia agriculture is a priority for the development of water resources. Since the establishment of the Water Supply Department under the Ministry of Animal Husbandry and Agriculture in 1938, until 1990 most attention was being given to the development of water supply for animal husbandry and agriculture. Industry development started in the middle of the 1930s. With the construction of apartment buildings, which started in 1960, a centralized water supply network was established. However, there was no integrated policy on water, and related organisations were established to settle only the immediate problems. The Water Supply Department of Ulaanbaatar was established in 1959.

An increase in the number of factories from the end of 1960 and intensification of the urbanisation were the social and economic prerequisites for the revision of the government policy on water. The agriculture sector was developing rapidly and the population concentration in urban and rural areas was increasing. At that time it was realized that the development of irrigated agriculture was a real necessity and this is also the situation today.

Measures taken during the years of intensive development of water resources resulted in major improvements of water supply for the population, industry, animal husbandry and agriculture. Water supply to the population in urban areas and industry was solved and many pastures were under irrigation. Apart from that, irrigation systems for watering crops and vegetable producing areas were set up and these have been producing about 20 percent of the required harvest.

Since the beginning of the implementation of irrigation the Russian Federation and other former socialist countries have assisted with the improvements in the water resources sector and with training of specialists.

The establishment of the Ministry of Water Economy of the Mongolian Peoples' Republic in 1965 was a significant factor in the further development of the water resources sector, including water supply for animal husbandry and crop and vegetable production.

Due to the dry character of the country, especially in the Gobi and steppe zones, a reliable harvest of all kind of vegetables and other crops is possible only using irrigation.

Because of the low precipitation (280–320 mm in the main irrigation areas, 90 mm in the Gobi and steppe zones), rainfed crop production is limited. As most precipitation falls in summer (80–90 percent) little humidity is kept in the soil. Most seasons are windy, especially the spring (April-May). Due to the low precipitation and windy conditions there is less possibility for snow to collect around plants to increase the soil humidity.

In the 1990s, 50 000 ha were irrigated. A simple irrigation system for 17 000 ha is now being used for crop production and the rest is used for pasture and hay production. At present Mongolia has no fish production in irrigation systems.

In 1986–1990, 645 000 ha of crops were planted annually, out of which 500 000 ha were for wheat, 11 200 ha for potatoes and 3 500 ha for vegetables, resulting in a harvest of 806 000 tonnes of crops, of which 633 000 tonnes was wheat, 127 000 tonnes potatoes and 46 000 tonnes vegetables. This satisfied the population demand for wheat flour, potatoes and vegetables. The surplus products were exported. As of 1990, 5.2 percent of the total agricultural area was irrigated.

But the crop production from irrigated areas was experiencing high operation costs, huge energy consumption and shortage of skilled and trained manpower on farms. It was not possible to fully exploit the production potential of irrigated areas. Due to technical and other reasons about 40 irrigated areas, covering a total of some 20 000 ha, were abandoned during the last 10 years.

All irrigation was done by overhead irrigation (spraying), which was difficult to operate and sometimes this method washed out the soil. It also had high operation and maintenance costs, and produced overcapacity which sometimes could not be harvested. As a result sometime more than 50 percent of the irrigation system was not used.

In the past Mongolia was self-sufficient in crop produce. In 1999, because of the hot weather conditions, there was a poor harvest. Irrigation was virtually forgotten although some was privatized, but individual farmers were unable to fully use it. The government has yet to take appropriate measures to revitalize the irrigated agriculture in the country.

**Table 1**

Agricultural area (in thousand hectares)

| Area, crops | 1989 | 1995 | 1996 | 1997 | 1998 |
|---|---|---|---|---|---|
| **Agricultural area, total** | **125 532.0** | **118 469.0** | **118 469.0** | **128 891.0** | **129 132.0** |
| out of which: | | | | | |
| arable land | 375.0 | 1 322.0 | 1 322.0 | 1 228.0 | 1 347.0 |
| meadows & pastures | 124 156.0 | 117 147.0 | 117 147.0 | 127 663.0 | 127 785.0 |
| irrigated land | 41 710.0 | – | 10 443.0 | – | 8 140.0 |
| **Sown area, total** | **837.9** | **372.6** | **347.8** | **333.8** | **326.6** |
| out of which: | | | | | |
| cereals | 673.4 | 356.5 | 332.6 | 316.9 | 306.9 |
| potatoes | 12.6 | 6.6 | 6.9 | 6.6 | 8.1 |
| vegetables | 4.2 | 3.2 | 3.2 | 4.3 | 5.5 |
| fodder crops | 147.1 | 6.0 | 4.3 | 4.7 | 4.9 |

Table 1 shows that from the total agriculture area in 1989, 4.9 percent was irrigated, in 1996, 3.03 percent, in 1998, 2.4 percent. It also shows that the irrigated areas have been steadily decreasing.

In 1986, 869 300 tonnes of crops, out of which 15 700 tonnes (2 percent) were harvested from irrigated areas, can be compared with the harvest in 1999 of 171 200 tonnes of crops, out of which 2 500 tonnes (or 1.4 percent) were harvested from the irrigated areas. In other words, over a period of 15 years there was a steady decline to only one fifth of the original production (by tonnage). Comparing 1999 with 1986 the production of wheat declined 3.9 times, potatoes 1.08 times, vegetables 1.1 times, planted fodder crops 40 times. Irrigation systems larger than 500 ha are now privatized, but the privatization has yet to show results in crop production. The programme of rehabilitation of agriculture includes the provision of a soft loan but there has been no investment given for the rehabilitation of irrigation systems.

**Table 2**

Harvest from irrigated lands (1986–1999) (tonnes/ha)

| Year | 1986 | 1987 | 1988 | 1989 | 1990 | 1991 | 1992 | 1996 | 1998 | 1999 |
|---|---|---|---|---|---|---|---|---|---|---|
| All crops | 1.4 | 1.1 | 1.3 | 1.3 | 1.1 | 1.0 | 0.8 | 0.7 | 0.6 | 0.6 |
| Wheat | 1.4 | 1.2 | 1.4 | 1.3 | 1.1 | 1.0 | 0.9 | 0.7 | 0.6 | 0.6 |
| Vegetables | 12.2 | 11.9 | 13.8 | 14.1 | 11.4 | 8.5 | 7.4 | 7.5.5 | 8.4 | 8.1 |
| Other | 3.7 | 3.8 | 3.1 | 3.7 | 4.5 | 2.7 | 2.6 | 4.4 | 3.0 | – |
| Irrigated fruitberries | 0.6 | 0.5 | 0.4 | 0.4 | 0.8 | 0.6 | 1.2 | 0.3 | 1.4 | 1.2 |

Table 2 shows a decrease in crop production on irrigated lands.

It should be noted that in Mongolia it is more convenient to use watering equipment with a small capacity, and it is expected that farmers know about soil humidity and some other basic principles of irrigation farming. Training in irrigation is essential if progress is to be made.

## 3. FISH FAUNA AND FISHERIES IN MONGOLIA

### 3.1 Background

Mongolia has 75 species of fish which belong to 36 genera and 11 families. There are over 30 species of commercial value. In Mongolia, main food components have traditionally been meat and dairy products and only foreigners residing in Mongolia have eaten fish. An increasing number of nationals who have experienced life abroad now also eat fish. Eating habits are changing especially in urban areas such as Ulaanbaatar. Demand for fish is rising and the need for a better utilisation of fishery resources is recognized.

In 1991, the Government of Mongolia outlined the policies for the development of fisheries:

- Management and exploitation of fishery resources in lakes and rivers of Mongolia
- Aquaculture
- Stocking of reservoirs.

Several studies and experiments have been carried out by three national institutes but implementation of the results and commercial fishery development have yet to take place largely because of lack of experience.

### 3.2 Fish species distribution

There are three major drainage basins in the country, i.e. the Arctic Ocean Drainage basin, the Pacific Ocean Drainage basin, and the Central Asian Inland Basin. Of the over 4 000 lakes of Mongolia, commercial fish stocks are present in Tsagaan Lake and Terkhiin Tsagaan Lake, located in the Arctic Ocean Drainage Basin. Tsagaan Lake, situated in the Darkhad depression in northern Mongolia, has an estimated annual fish production of approximately 100 to 150 tonnes, while that of Terkhiin Tsagaan Lake is 50 to 100 tonnes.

The fish production capacity of Buir Lake in the Pacific Drainage Basin is around 200 tonnes per year. The combined sustainable fish production of all lakes in the Central Asian Inland Drainage Basin is estimated at more than 300 tonnes.

For additional information on fish and fisheries of Mongolia see Dulmaa (1999).

### 3.3 Preservation of endangered fish species

The Mongolian Law on Hunting adopted in 1995 and supported by other laws and regulations has played an important role in the development of fish culture as well as in the proper use and conservation of fish species in the country.

According to the provisions of this law, the Amur sturgeon (*Acipenser schrenki*) and Siberian sturgeon (*Acipenser baeri*) are listed as protected species while grass carp (*Ctenopharyngodon idella*), silver carp (*Hypophthalmichthys molitrix*), and tench (*Tinca tinca*) are registered as endangered. In addition, six species of fish have been included in the Mongolian Red Book and measures have been taken to conserve these species.

In order to preserve the fish stocks on a large scale and to maintain favourable conditions for their breeding, several lakes such as the Khuvsgul, Uvs and Tsagaan and important parts of some rivers have been categorized as protected areas.

At present there are cases of illegal fishing, an activity that employs the use of harmful methods. This has especially posed threats to taimen (*Hucho taimen*) and Siberian whitefish stocks in lakes.

In Naiman Lake in the Uvurkhangai Province of Central Mongolia, 25 000 and 50 000 fry of the omul (*Coregonus autumnalis*) and peled whitefish (*Coregonus peled*) were stocked in 1978 and 1979, respectively. Between 1983 and 1986, around 10 million omul fry were imported from a fish hatchery in the former USSR and introduced successfully into a group of lakes in Bayan Ulgi Province in Western Mongolia. In 1986, millions of Baikal omul fry from the former USSR were also introduced into Lake Tsagaan in Khuvsgul Province.

Between 1991 and 1995, research into propagation and conservation of taimen and Siberian grayling (*Thymallus arcticus*) was conducted in the lakes of the Darkhad Depression.

### 3.4 Development of fishing industry in Mongolia

Large scale fisheries in Mongolia started in the 1950s and targeted lakes Buir (located in the far eastern edge of Mongolia), Ugii Nuur (central Mongolia), and Tsagaan Nuur (northern Mongolia).

The confirmed annual catches from these three lakes peaked at 779.1 tonnes in 1959. Afterwards, partly due to overfishing, the catches never exceeded 400 tonnes, and until the 1990s fish catches from all water bodies in Mongolia ranged from 120 to 200 tonnes per year.

Fish are captured seasonally as during the cold season they can be preserved by freezing under the prevailing very low temperatures. This makes possible their transport to markets. During the warm period of the year fish cannot be preserved as yet, hence there is no fishing and fishermen have only seasonal jobs. The lack of preservation possibilities during the warm season interrupts the supply of fish to markets.

Due to the collapse of the system of centrally planned economy in 1990, fisheries and other economic sectors became disorganized and lost their previous operational efficiencies. As a result there has been a sharp reduction in catches.

Traditionally, in Mongolia main food policies center on agriculture, especially on livestock production. However, demand for fish products is rising not only because of the presence of the resident foreigners and some 10 000 tourists visiting Mongolia every year, but also because Mongolians are demanding a greater variety of food. Thus there is an urgent need for the restoration and further development of fisheries in Mongolia, a landlocked country with no access to ocean fishery resources.

As the natural resources in lakes and rivers are limited and easily overexploited there is a need to develop fishery as a new resource of food. More attention has to be paid to the aquaculture potential of the country, as well as to introducing better preservation and processing methods of fish produced through capture fisheries. Better control and management of natural fishery resources is essential for protection of the fish stocks so that they can be harvested at a sustainable level.

In order to restore and strengthen national capacity in fishery management and development in Mongolia the following priority areas have been identified:

- Need for a balanced development of fishery at sustainable level
- Protection of breeding stocks
- Hatchery production of stocking material
- Pilot studies of aquaculture potential under Mongolian conditions
- Improved collaboration between fishermen and fishery scientists
- Better fish processing and a diversification of fishery products to satisfy especially the tourism demand
- Promotion of rural economy.

## REFERENCES

Dulmaa, A. 1999. Fish and fisheries in Mongolia. FAO Fisheries Technical Paper. No. 385: 187–236. FAO, Rome.

# FISHERIES DEVELOPMENT IN XINJIANG, CHINA

by

**Guo Yan**
**Fishery Research Institute of the Xinjiang Uygur Autonomous Region,**
**Urumqi, 830 000 China**

## ABSTRACT

Xinjiang is located in the hinterland of Eurasia. It is the largest province in China, covering an area of 1.6 million km$^2$ and accounting for one sixth of China's territory. Prior to the 1950s Xinjiang had almost no fishery. Since then, with the economic development, its fishery has made enormous progress. In 2000 the output of the fishery in Xinjiang reached 58 835 tonnes, of which aquaculture produced 49 761 tonnes, i.e. 64.6 percent of the total. Capture fisheries and aquaculture include reservoir fisheries, net-cage culture, culture in warm water and cold water. Fish species captured and/or cultured are common carp, silver carp, bighead carp, golden carp, grass carp, *Megalobrama amblycephala*, rainbow trout, tilapia, *Clarias leather* and some other fish. Crabs, soft-shelled turtles and shrimps are also cultured. As the fisheries are still developing, many problems are emerging, such as lowering of the ground water level, eutrophication, disappearance of the local fish species, spread of fish diseases. The increasing rate of water diversion and water consumption for irrigation and other purposes is partly responsible for this.

## 1. BACKGROUND

The Xinjiang Uygur Autonomous Region is situated in northwestern China, adjacent to Gansu and Qinghai provinces in the east, the Tibet Autonomous Region in the south, with Pakistan, India and Afghanistan in the southwest, Tajikistan, Kyrgyzstan and Kazakhstan in the west, and Russia and Mongolia in the north and northeast. Xinjiang is equally distant from both the Pacific and the Atlantic oceans, i.e. about 4 000 km from each. Xinjiang covers 1.6 million km$^2$ and represents one sixth of the territory of China. It has a population of 18 million, mainly Uygurs, Han, Mongols, Kazakhs and Hui.

Xinjiang is surrounded by high mountains: in the north and northeast are the Altay Mountains, in the northwest the Tarbagatai Mountains, in the west and southwest the Pamir and Karakoram Mountains, in the south the Kunlun Mountains, and in its central section are the Tien Shan Mountains crossing it from east to west and dividing Xinjiang into two parts, the Tarim Basin in the south and the Dzhungarian Basin in the north. The average altitude of Xinjiang is about 1 000 m above sea level, with the highest mountains of Tien Shan and Kunlun reaching over 6 000 m. In the middle Tarim Basin the elevation is 800–1 300 m, in the Dzhungarian Basin 500–1 000 m. Lake Aydingkol in the Turpan Basin is the lowest point at 155 m below sea level. Sixty percent of Xinjiang is covered by desert, 30 percent by steppes, and the rest is oases, forests, agricultural areas, lakes and other waterbodies.

Xinjiang is an arid region. In summer it is very hot, in winter it is very cold. There are differences of up to 20 °C between the night and day temperatures. The highest average July temperature in northern Xinjiang is 25 °C, in southern Xinjiang 28 °C. The lowest temperatures are reached in January, down to –18 °C in the north, –9 °C in the south. Xinjiang has the highest number of sunshine hours per year in China, with most areas receiving 2 900 hours. The annual solar radiation is estimated at 130–140 kcal/cm$^2$.

Dry weather is a characteristic feature of Xinjiang. The average annual rainfall is only 150 mm, and more than half of the plains receive less than 100 mm. It rains more in the north than in the south, and more in the west than in the east. The south receives 30–50 mm, the west and north 40–70 mm, the east 20–40 mm, and the Altay and Tien Shan mountains 100 mm. The north and west have a lower evaporation rate than the south and east. The average annual evaporation differs, is lowest in the north and northwest (1 600–2 000 mm), and highest in the east (2 800–3 000 mm).

There are 570 rivers in Xinjiang, of which 144 are in southern Xinjiang, and 426 in northern Xinjiang. They originate from the northern and southern slopes of Tien Shan, western slopes of Altay, northern slopes of Kunlun, and eastern slopes of the Pamir. They are fed by ice and snowmelt and by rainfall precipitation. There are also 139 lakes larger than one square kilometer each, covering a total of 513 000 ha, and 479 reservoirs.

## 2. XINJIANG FISHERIES

Prior to 1958 fish production was limited and on a small scale. Subsistence fishermen used primitive gear to capture a few fish for their families. It was the farmers and herdsmen who were mainly fishing, using forks and hooks. There was no aquaculture at that time. This was followed by a period of fishery development from 1958 until 1971. This was a gradual process, which included the start of aquaculture, induced breeding, and fish species introductions. The fishing gear became more sophisticated and fishery was better managed. Since 1972 there has been a rapid expansion of fish pond culture, use of formulated fish feed, and intensification of aquaculture on a large scale. Reservoir fisheries enhancement has included introduction of fish culture methods and fertilisation of some reservoirs to increase their productivity. Fish producers have been trained in new fish culture technologies. New species have been introduced in aquaculture which has led to an increase in income. There have been improvements in fish processing as well. Fisheries laws and regulations have been promulgated and enforced.

The present capture fisheries concentrate predominantly on two lakes: Lake Bosten and Lake Ulungur, and three rivers: Ertix = Irtysh, Yili=Ili, and Tarim. Prior to 1980 the fish captured from these waters represented 60 to 80 percent of the total Xinjiang fish production. Lake Ulungur covers 100 000 ha. The following fish have market importance: the indigenous species *Perca fluviatilis* (perch), *Leuciscus leuciscus* (dace), *Rutilus rutilus* (roach), *Carassius auratus gibelio* (golden carp), and the introduced species *Hypophthalmichthys molitrix* (silver carp), *Aristichthys nobilis* (bighead carp), *Abramis brama orientalis* (bream) and *Hypomesus olidus*. During the period 1958–1999 there were two exceptional peaks in fish production from this lake: in 1976 – 4 500 tonnes, and in 1999 – 5 100 tonnes, but mostly the fish production was around 3 000 tonnes per year.

Lake Bosten covers an area of 90 000 ha, and has a salinity of 1.5 g/litre. The average annual fish catch is 2 500 tonnes. The following fish species have market value: endemic species *Aspiorhynchus laticeps*, *Schizothorax* (*Racoma*) *biddulphi*, and the introduced silver carp, bighead carp, golden carp, grass carp (*Ctenopharyngodon idella*), *Hypomesus olidus* and several other species. The indigenous fish have disappeared from catches, *A. laticeps* disappeared by 1974, *S. (R.) biddulphi* by 1985. All three species of Chinese carps and *Hypomesus olidus* were firmly established in catches by 1994. Common carp, golden carp, perch and dace were a regular component in all catches during the 1972–1999 period. During the same period there was a gradual increase in fish catches from 200 tonnes in 1972 to 2 000 tonnes in 1979, followed by a slow decline to 1 700 tonnes in 1985, which was followed by an increase to 3 000 tonnes in 1988, eventually reaching a maximum of 3 500 tonnes in 1996 and a decline to 3 300 tonnes by 2000.

The Ertix (Irtysh) River drains north and eventually, after passing through Kazakhstan and Russia, it enters the Arctic Ocean. The annual fish catch ranges between 200 and 400 tonnes. The major fish species captured are *Abramis brama orientalis*, *Esox lucius* (pike), *Leuciscus idus*, *L. leuciscus baikalensis*, *Rutilus rutilus*, common carp, and perch. In the Ili River, which flows west, entering Kazakhstan and eventually ending in Lake Balkhash, the major fish species captured prior to the 1960s were *Schizothorax (Racoma) argentatus* and S. (*R.*) *pseudaksaiensis*, but these soon disappeared and were replaced by *Abramis brama orientalis*, *Aspius aspius*, common carp, *Silurus glanis* (wels), grass carp, silver carp, and several other species. The annual fish catch ranges from 100 to 300 tonnes. The Tarim River is the largest river in Xinjiang. Again, the initially fished indigenous species *Aspiorhyncus laticeps* and *S. (R.) biddulphi* soon disappeared and were replaced by the introduced common carp, grass carp, silver carp and bighead carp. They have been supporting fisheries of up to 200 tonnes annually.

The first aquaculture farms were established in the 1960s. At present ten of them work well and their production potential is 300 million fry per year. In the year 2000 there were 74 state-owned enterprises and 3 158 households engaging in aquaculture production. The total number of fishermen was 18 177. The following fish were produced under aquaculture conditions: common carp, golden carp, silver carp, *Megalobrama amblycephala*, *Clarias leather*, rainbow trout, tilapias, and also freshwater shrimps and crabs.

For aquaculture production in Xinjiang in year 2000 see Table 1.

**Table 1**
Aquaculture production in Xinjiang by type of waterbody

| | Ponds | Lakes | Reservoirs | Pit ponds | Other |
|---|---|---|---|---|---|
| Area (ha) | 10 384 | 4 851 | 43 735 | 10 416 | 469 |
| Production (tonnes) | 36 966 | 639 | 6 965 | 2 493 | 867 |
| Yield (kg/ha) | 3 560 | 132 | 205 | 234 | 1 849 |

Table 1 indicates that most of the aquaculture production comes from fish ponds, followed by reservoirs, pit ponds, other types of waterbodies, and the smallest amount comes from aquaculture in lakes. The highest yields are achieved in ponds, which also produce the highest percentage of fish from all types of waterbodies; the lowest yields and percentage production of the total come from lakes.

In 2000 the total fish production in Xinjiang was 58 835 tonnes, of which aquaculture produced 49 761 tonnes (84.6 percent) and capture fisheries 9 074 tonnes (15.4 percent). The state-owned fishery produced 43.6 percent and the private fishery 56.4 percent.

For additional information on Xinjiang fish and fisheries, published in English, the reader may wish to consult Walker and Yang (1999).

## 3. CURRENT PROBLEMS OF FISHERIES IN XINJIANG

For an arid region such as Xinjiang water is the most essential resource for human activities. During the last 50 years the population of Xinjiang has experienced fast growth, and the demand for water has increased accordingly. The area of irrigated land has increased, there are new industries, and the increasing population needs more water as well. As a result of the increasing uptake and diversion of water the lower courses of rivers have been receiving less water, and some sections have desiccated completely. The level of ground water has been declining. Terminal lakes are receiving less water or no water at all as the river and stream water does not reach them any more. Soils are getting more alkaline, and water more saline. The large scale irrigation efforts have destroyed the environment of the original fish fauna, and as a consequence indigenous fish species have gradually disappeared; overfishing has also contributed to this. The following are some of the major impacts of changes in the use and redistribution of water.

The original fish fauna of the Tarim River and Lake Bosten had the valuable species *Aspiorhynchus laticeps* and *Schizothorax (R.) biddulphi*. In spite of that during the 1960s additional fish species were introduced from the Chang Jiang (=Yangtze) and Ertix rivers. These introductions, combined with high fishing pressure and water resource development projects, led to a rapid decline in stocks of both species. The species are now listed as protected. The same happened in the Ili River with *Schizothorax. (R.) argentatus* and *S. (R.) pseudaksaiensis*. In the Ertix River three other species were impacted in the same way: *Stenodus leuciscus nelma*, *Acipenser ruthenus* and *A. baeri*. These three species and *S. (R.) argentatus* are now believed to be extinct.

During the 1970s there was a rapid development of fish ponds, and by the early 1990s they covered 10 000 ha. Most fish ponds were constructed close to cities and supplied with ground water. This has had a negative impact on the level of ground water, which not only will result in shortage of water for aquaculture in the future, but is already affecting agriculture production.

The introduction of semi-intensive and high density aquaculture, and the use of fertilizers, fish feed and net-cages in the reservoirs Tajiaowan, Luigong, Beijiantan, Daquangou and Moguhu, increased the yields to 3 000 kg/ha, but water quality has deteriorated along with the accumulating organic matter on the bottom of these reservoirs. To avoid fish mortalities, it has been necessary to reduce the density of fish in these cultures.

Other problems have also appeared, such as reduction in the genetic quality of the cultured fish stocks, deterioration of fish feed quality, and increasing occurrence of fish diseases, probably due partly to the previous two factors. There has been an increase in fish mortalities especially during winter. The continuous production of common carp, golden carp and silver carp in the same ponds might also have led to the increased mortalities. What is becoming more and more clear is that there is a need not only for improving the quality of the cyprinid fish produced for markets, but also for culturing fish species of a higher quality, such as non-cyprinids.

## 4. DISCUSSION

Future fisheries development in Xinjiang will have to consider the compatibility of the intensified aquaculture with the rising demands for water resources. Overexploitation of indigenous fish species, which in some cases has led to their local extinction, followed by aquaculture using introduced species, has led to further problems. These include the deterioration of the aquatic environment due to the intensification of fish production, diminishing of water resources available for aquaculture, especially groundwater supply, impacts of other uses of water such as irrigated agriculture, municipal and industrial water supply, which cause water diversion and sometimes water pollution, and also may lead to desiccation of the lower courses of rivers and terminal lakes.

Further development of aquaculture and enhanced fisheries may depend on the identification and production of better quality fish. Water-saving aquaculture methods need to be tested and introduced in all available suitable waterbodies. The use of fertilizers, both organic and inorganic, should be more judicious, and water quality should be improved through planting aquatic macrophytes and other measures. If fish species introductions are still to be considered, strict quarantine measures should be maintained before any introduction is made. Better fish strains are needed for aquaculture, which has been using the same fish stocks for many years. There is also need to test new methods and methodologies of fish production and improve fish processing. Introducing better quality fish species into culture will result in a better income for fish farmers.

## REFERENCES

Walker, K.F. & Yang, H.Z. 1999. Fish and fisheries in Western China. *In*: Fish and fisheries at higher altitudes: Asia (Petr, T. ed.). FAO Fisheries Technical Paper. No. 385: 237–278.

# THE USE OF IRRIGATION SYSTEMS FOR SUSTAINABLE PRODUCTION OF AGRICULTURAL AND FISH PRODUCTS IN THE REPUBLIC OF KAZAKHSTAN

**by**

**Khismet Ismukhanov**
**Kazakh Scientific and Research Institute of Fisheries, 480016.89[A], Suinbai Str., Almaty, Kazakhstan**
**and**
**Valiakhmet Mukhamedzhanov**
**Kazakh Scientific and Research Institute of Aquaculture, 484022.12, Kolbasshi Str., Taraz, Kazakhstan**

## ABSTRACT

Kazakhstan has 8 large river basins. The total annual water discharge in Kazakhstan is 100.9 km$^3$ of which 56 km$^3$ originates in the country itself, with the rest coming from the neighbouring countries. The available water resources (minus those needed for maintaining good water quality and for ecological needs) represent 42.6 km$^3$ per year. Of the annual use of 42.6 km$^3$ of water over 70 percent is used for irrigation. Irrigated farming is essential for food production in Kazakhstan. In Kazakhstan approximately one third of agricultural plant production comes from the irrigated lands which represent about 5–6 percent of the farmed area. Irrigated farming predominates in the south of the country where cotton is produced on 36.6 percent (i.e. 2.3 million ha [mha]) of the irrigated area, forage crops on 17.8 percent, cereals on 14.5 percent, vegetables and other cultures on 29.1 percent. To store water for irrigation, many rivers, in particular the large ones, are regulated by dams. This allows both long-term and seasonal regulation. There are 475 reservoirs in Kazakhstan and more than 96 000 km of irrigation canals at present. In the southern regions of Kazakhstan there are 75 reservoirs with a total capacity of 95.5 km$^3$ and surface area of over 10 000 km$^2$. The great majority of large reservoirs are multipurpose. Reservoirs play an important role in fisheries. Among the large reservoirs, fisheries are well developed on the Bukhtarma in the north, and on Chardara and Bugun in the south. Thirteen fish hatcheries-cum-fish farms have been constructed and currently operate in Kazakhstan. Most of them are adjacent to reservoirs, including Bukhtarma and Kapshagay, and they produce fish for stocking the reservoirs. They have played a significant role in enhancing stocks of common carp and silver carp. Lately, economic problems in the country have led to a deterioration of irrigation systems, including pumping stations, and the undergoing process of transition to market economy has interfered with the formerly efficient and well-organized fisheries. As a result, the catches have been declining, fish stocks are often harvested illegally and reservoir-associated hatcheries are facing shortages of funds and cutting down on the number of stocking material produced. The great fisheries potential of numerous waterbodies is thus poorly utilized at present.

# 1. IRRIGATION FARMING SYSTEMS IN KAZAKHSTAN

## 1.1 Water resources

Irrigated farming is essential for food production in Kazakhstan. Prior to transition to the market economy approximately one third of agricultural products came from irrigated lands, although this represented only 5–6 percent of the farmed area. In southern Kazakhstan the produce from irrigated farming often represented 2/3 of the total produced in Kazakhstan. Over 70 percent of water is used for irrigation. At the same time the efficiency of production per irrigated hectare does not reach the optimum level.

The efficiency of irrigation farming and its further development in Kazakhstan are determined mostly by the availability and quality of water resources. Rational use of water resources is a top priority. The ever growing demand for water and the continuing deterioration of water quality resulting from pollution of all kinds require urgent and efficient measures to safeguard sustainable development of agriculture and fisheries and other important sectors of the national economy.

While areas suitable for irrigated agriculture cover about 5 mha of land, water resources of Kazakhstan, especially in the south, are limited. There are 8 large river basins and more than 96 000 km of irrigation canals. The total annual water discharge in Kazakhstan is 100.9 km$^3$ of which 56 km$^3$ originates in the country itself, with the rest coming from the neighbouring countries. The available water resources (minus those needed for maintaining good water quality and for ecological needs) represent 42.6 km$^3$ per year (Table 1).

**Table 1**

Water resources of Kazakhstan

| Major rivers | Average annual discharge(km$^3$) | | Available water resources under different levels of supply (%) | | Current use of water (km$^3$) |
|---|---|---|---|---|---|
| | Total | From neighbouring countries | 75 | 95 | |
| Total in Kazakhstan, of which | 110.9 | 43.9 | 32.6 | 26.0 | 36.6 |
| Syr-Darya | 17.9 | 14.2 | 9.8 | 9.3 | 12.0 |
| Ili | 17.8 | 11.1 | 4.3 | 3.4 | 5.2 |
| Karatal, Lepsy, Ayaguz etc. | 10.0 | 0.3 | 2.7 | 2.0 | 3.6 |
| Irtysh | 33.8 | 7.8 | 10.8 | 8.4 | 5.4 |
| Ishym | 2.3 | – | 0.4 | 0.1 | 1.0 |
| Nura | 0.8 | – | 0.2 | – | 1.1 |
| Sarysu | 0.4 | – | 0.1 | – | 0.4 |
| Tobol | 0.6 | – | 0.1 | – | 0.5 |
| Torga | 1.4 | – | 0.2 | – | 0.5 |
| Shu | 282.4 | – | 2.0 | 1.6 | 2.7 |
| Talas-Assa | 1.4 | 0.7 | 1.0 | 0.7 | 1.4 |
| Ural | 9.5 | 6.5 | 0.6 | 0.3 | 2.0 |
| Emba, Sagiz | 0.8 | – | 0.2 | – | 0.2 |

The limited water resources place constraint on further development of agriculture, and shortage of water resources in drought years leads to restricted water supply in some parts of the country. This takes place when water resources of most river basins in the country are either exhausted (basins of the rivers Syr-Darya, Shu, Talas, Ili, etc.) or almost exhausted.

The shortage of water resources leads to the use of return waters for irrigation. Under "return waters" is understood the part of irrigation water which returns from irrigated land back to the river or is discharged into a depression or a lake without an outflow. Return waters result in degradation of flora and fauna and in pollution of surface water resources as they contain higher concentrations of salts, fertilizers, herbicides and other harmful chemicals, as well as higher concentration of bacteria than water present in unpolluted rivers. An example of this is the deterioration of water quality in the Syr-Darya River under the impact of return waters.

To accurately estimate the quantity of return waters entering a river is not easy. During the period 1986–1995 an average annual volume of the non-return water resources used in the Syr-Darya River basin was estimated at 23.86 $km^3$ per year, with annual fluctuations from 18.79 to 24.27 $km^3$ per year. In Kazakhstan the average value was 6.28 $km^3$ per year.

In the Syr-Darya River basin 8 160 $m^3$/ha of freshwater is used for irrigation. In the upper reaches of the river (Kyrgyzstan, Uzbekistan) it is less, with only 6 320 $m^3$/ha used (range 5 760–7 170 $m^3$/ha). The maximum is used in the middle course and the lower course of the river (in Kazakhstan), with a range from 8 290 $m^3$/ha to 13 560 $m^3$/ha.

Poor technical conditions of Syr-Darya irrigation systems necessitate the use of return water, of which currently over 10 $km^3$ is used, including the 3.2 $km^3$ formed within Kazakhstan. Large volumes of return water were used especially during 1975–1997. This has been followed by a slow decrease as a result of the reduction of the area of irrigated lands and consequently also the volume of irrigation water required. The neighbouring countries Uzbekistan and Kyrgyzstan continue to produce considerable volumes of return water which is being discharged into the Syr-Darya and reaches Kazakhstan. This results in the deterioration of the quality of irrigation water and aggravates the ecological and epidemiological situation in the Syr-Darya catchment.

The use of return water and freshwater (non-return water) in irrigation systems of southern Kazakhstan and in Kyzylordinsky Oblast (=Province) during 1990–2000 is shown in Table 2.

Many rivers, in particular the large ones, are regulated by dams. This allows both long-term and seasonal regulation. There are 475 reservoirs in Kazakhstan at present. The characteristics of the major reservoirs are given in Table 3. In the southern regions of Kazakhstan there are 75 reservoirs with a total capacity of 95.5 $km^3$ and surface area of over 10 000 $km^2$. This includes 22 reservoirs of long-term regulation. The capacity of 11 reservoirs exceeds 100 million $m^3$ ($Mm^3$) . The great majority of reservoirs are multipurpose. In eastern and northern Kazakhstan their first priority is hydropower production, in northern Kazakhstan they also provide water supply for cities and towns and for industries and support fisheries, in the south they serve mainly irrigation and fisheries, and in western Kazakhstan in the area between the Ural and the Bolshoi Uzen rivers they serve irrigation.

## 1.2 Irrigation farming

Irrigation farming is the main consumer of water. Its share is more than 70 percent of the total water intake. Some decline in water consumption has been observed. Compared with 1990 the total water consumption has decreased by 42.6 percent, with a 33.9 percent decrease in agriculture.

**Table 2**

**Average annual volumes of the use of non-return and return water resources in irrigation systems of the Syr-Darya River basin in Kazakhstan**

| Irrigation system | Area of irrigated lands ('000 ha) | Actually irrigated, ('000 ha) | Volume of water intake (Mm$^3$) | Volume of return waters (Mm$^3$) | Volume of the use of non-return water resources, (Mm$^3$/year) |
|---|---|---|---|---|---|
| | | South-Kazakhstan Oblast | | | |
| Mc "Dostyk" system | 125 320 | 123 800 | 66 070 | 211 222 | 448 848 |
| CHAKIR | 65 553 | 57 925 | 478 939 | 153 260 | 325 679 |
| Kyzylkumsky large tract | 76 043 | 66 100 | 733 741 | 243 797 | 498 944 |
| ARTHUR | 203 680 | 193 890 | 1 281 140 | 409 965 | 871 175 |
| incl.the Arys River basin | 108 000 | 102 801 | 823 300 | 263 530 | 560 000 |
| Shauldersky system | 29 280 | 27 872 | – | – | – |
| Arys-Turkestansky system | 66 400 | 63 217 | 457 610 | 146 435 | 311 175 |
| Total in the South-Kazakhstan Oblast | 470 596 | 441 715 | 3 153 890 | 1 009 244 | 2 144 646 |
| | | Kyzylordinsky Oblast | | | |
| Toguskensky large tract | 31 842 | 25 810 | 414 036 | 144 913 | 269 123 |
| Zhanakorgano-Shiiliysky system | 45 454 | 36 840 | 591 031 | 206 861 | 384 170 |
| Right-bank Kyzylordinsky | 26 437 | 21 427 | 343 756 | 120 315 | 223 441 |
| Left-bank Kyzylordinsky | 88 449 | 71 687 | 1 150 089 | 402 531 | 747 558 |
| Right-bank Kazalinsky | 16 440 | 13 324 | 158 453 | 55 459 | 102 994 |
| Left-bank Kazalinsky | 20 637 | 16 726 | 267 859 | 93 751 | 174 108 |
| Total in the Kyzylordinsky Oblast | 229 259 | 185 810 | 2 925 224 | 1 023 830 | 1 091 394 |
| Total in the Syr-Darya River | 699 855 | 627 525 | 6 079 114 | 2 033 074 | 3 236 040 |

Irrigation farming in Kazakhstan, especially in its southern regions, has come a long way, from the primitive irrigation ditches, aryks, to the most complex modern engineering systems with elements of automatization. By the beginning of the 1990s the area of irrigated lands alone exceeded 2.3 mha. Improvement funds and the share of products from irrigated areas reached almost one third of the gross production output of farming.

**Table 3**

Characteristics of the major reservoirs of Kazakhstan

| Economic regions | Storage reservoirs | Rivers | Started operating | Volume $Mm^3$ | Area $km^2$ | Salinity g/L |
|---|---|---|---|---|---|---|
| Sourthern | Chardara | Syr-Darya | 1965 | 5 700 | 900 | 1.3–1.7 |
| | Kapshagay | Ili | 1970 | 23 100 | 1 847 | 0.4–0.6 |
| | Tashutkul | Shu(=Chu) | 1974 | 620 | 78 | 0.6 |
| Eastern | Ustkamenogorsk | Irtysh | 1952 | 600 | 37 | 0.13–0.16 |
| | Bukhtarma | Irtysh | 1960 | 49 600 | 5500 | 0.12–0.14 |
| Northern | Verkhneusol | Tobol | 1971 | 816 | 87 | 0.6–1.1 |

In many drought-prone regions of Kazakhstan it is not possible to maintain agricultural production without irrigation. Being one of the main directions of land melioration, irrigation permits land productivity to increase significantly. Irrigation is especially effective in southern Kazakhstan. It is known that irrigation increases crop yields 3–7 times compared to those produced on non-irrigated areas in the desert and semi-desert climate. Irrigation farming makes it possible to grow higher value crops such as sugar beet, cotton, rice, vegetables, melons, fruits and berries. By percentage cotton, with 38.6 percent of the total, dominates the crops produced using irrigation in southern Kazakhstan, followed by animal feed crops (17.8 percent), cereals (14.5 percent), vegetables (9.5 percent), and other crops (19.6 percent). Rice is cultivated on about 70 000 ha in three Oblasts of Kazakhstan.

Irrigation farming reached a maximum surface area in 1986 with regular irrigation of 2.4 mha. At that time 1.034 mha were irrigated in the estuary of the Syr-Darya. After that the productivity of irrigated lands decreased (Table 4).

**Table 4**

Crop yields on irrigated lands in southern Kazakhstan (tonnes/ha)

| Crops | 1986 | 1990 | 2000 |
|---|---|---|---|
| Total cereals (average) | 3.3 | 3.5 | 2.6 |
| Cotton | 2.6 | 2.7 | 1.9 |
| Sugar beet | 27.8 | 25.4 | 14.6 |
| Vegetables (average) | 18.9 | 17.8 | 13.7 |
| Potatoes | 12.3 | 10.1 | 6.4 |
| Perennial herbs for hay | 5.1 | 6.8 | 5.7 |

At present the area of irrigated lands totals 2.36 mha. Only 1.38 mha are in a good condition, i.e. about 60 percent of irrigated lands. The collector-drainage system is used on a small part of irrigated lands. Lands are irrigated mainly by a surface method. There is a low level of mechanization. Overhead irrigation is used on an area of about 660 000 ha. There are 96 387 km of irrigation canals, of which 20 620 km are lined, and 9 467 km are pipelines. The length of the collector–drainage system is 14 902 km. In recent years there has been no further development of the irrigation agriculture, rather deterioration of the existing structures due to the virtual absence of government financial support. This reflects the deterioration of the overall economic situation in agriculture. Almost one million hectares of irrigated lands are now out of production, there has been an increase in poor practices of water use and construction and rehabilitation have been suspended. As a result, irrigation systems as well as the quality of lands are gradually deteriorating.

The cultivated irrigated lands have been reduced by 880 000 ha, i.e. 43.4 percent, over the last five years. There have also been considerable changes in crop production. By 1997 cultivation of livestock feed crops had been reduced by more than 2.7 times; sowing areas of vegetables, melons and potatoes by 2.2 times, and non-food crops by more than one third. Where formerly cereals were grown on 30 percent of irrigated areas, today they are produced on almost 50 percent of them.

Changes in irrigation over the period 1970–1990 in Kazakhstan are shown in Tables 5 and 6.

**Table 5**

Changes in irrigation farming in Kazakhstan

| Indices | Years | | |
|---|---|---|---|
| | 1970 | 1988 | 1998 |
| Total area of irrigated lands, thous. ha,<br>of which under agriculture: | 1 370.9<br>1 248.0 | 2 321.9<br>2 235.0 | 2 361.8<br>1 978.0 |
| Total irrigated area of arable lands, thous. ha,<br>of which overhead irrigation: | 1 188.1<br>135.0 | 2 202.7<br>664.2 | 1 767.4<br>210.3 |
| Unused (not irrigated) areas requiring irrigation, total, thous. ha | 137.1 | 119.2 | 594.4 |
| the same, in % of the total area: | 10.3 | 5.1 | 25.2 |
| Water intake from sources of irrigation, bln. $m^3$ | 17.3 | 23..3 | 16.2 |
| Water supply (farms), bln $m^3$ | 13.5 | 18.9 | 12.6 |
| Coefficient of efficiency | 0.78 | 0.81 | 0.78 |
| Head water intakes in government owned systems (number) | 70 | 13.6 | 150 |
| Their repair cost, bln tenge | 34.8 | 96.6 | 99.7 |

The main irrigated areas are located in four southern Oblasts under drought conditions: Kyzylordinsky Oblast – 285 900 ha, South Kazakhstan Oblast – 500 500 ha, Zhambylsky Oblast – 248 200 ha, and Almaty Oblast – 665 900 ha. In total they occupy 1.7 mha out of the total of 2.2 mha of irrigated areas in all Kazakhstan. As for the other regions, the Eastern Kazakhstan Oblast has the largest irrigated tracts of lands of 220 000 ha.

Further expansion of irrigated lands faces two major constraints: the unfavourable social and economic situation in the country which is slowing down investment in the development of water resources, and chronical shortage of water in some areas. At present there exists no integrated management programme for the development of water and land resources for a number of river basins.

**Table 6**

Main indicators of the use of water resources in Kazakhstan, 1995–1998

| Indicator | 1995 | 1996 | 1997 | 1998 |
|---|---|---|---|---|
| Number of water users | 5 263 | 4 830 | 4 531 | 4 557 |
| Total freshwater intake, million $m^3$ | 28 807.1 | 26 482.8 | 24 977.8 | 22 317.0 |
| Including: surface water, million $m^3$ | 25 597.7 | 23 511.5 | 22 321.6 | 20 975.0 |
| ground water, million $m^3$ | 2 013.6 | 1 767.9 | 1 545.9 | 1 343.0 |
| Total use of freshwater, million $m^3$ | 22 239.2 | 20 495.2 | 18 337.6 | 16 004.8 |
| incl. for the following needs: | | | | |
| - drinking water | 1 241.6 | 1 141.4 | 826.3 | 736.9 |
| - industrial | 4 089.5 | 3 211.9 | 3 016.0 | 2 826.6 |
| - regular irrigation | 12 115.1 | 12 243.4 | 10 222.5 | 8 927.2 |
| - irrigation in river deltas | 3 679.8 | 2 936.4 | 3 421.0 | 3 001.8 |
| - agricultural water supply | 355.2 | 298.9 | 236.8 | 190.8 |
| - water supply development | 327.7 | 2 | 227.6 | 99.5 |
| - fish ponds | 319.9 | 242.1 | 169.7 | 98.8 |
| - other needs | 110.5 | 144.7 | 217.6 | 123.1 |
| Water disposal, total, million $m^3$ | 7 069.7 | 6 118.9 | 5 293.8 | 4 804.0 |
| incl. a) surface water units | 5 780.8 | 4 970.4 | 4 308.4 | 3 800.4 |
| - polluted | 229.8 | 190.6 | 188.2 | 168.8 |
| - of these without treatment | 26.6 | 27.0 | 23.4 | 24.7 |
| - clean water | 5 293.8 | 4 470.2 | 3 854.2 | 3 377.3 |
| - treated to required standard | 257.1 | 309.5 | 265.9 | 254.3 |
| b) underground disposal | 15.2 | 11.5 | 9.1 | 7.6 |
| c) into depressions | 1 273.6 | 1 137.0 | 976.3 | 995.6 |
| Upper limits of water intake, million $m^3$ | 32 464.4 | 31 209.1 | 35 172.9 | 31 220.0 |
| incl. from groundwater | 2 451.6 | 2 349.3 | 2 017.2 | 1 964.0 |
| Water re-use, million $m^3$ | 7 906.6 | 6 869.9 | 6 066.6 | 5 348.4 |

Government policy for the near future should first of all concentrate on the rehabilitation of irrigation systems but at the same time decisions should be made on how to optimally use water and land resources. This would allow the maintenance of the existing lands under irrigation.

Land ownership transfer from the state to private ownership can be considered as completed. Further work in this direction should deal with structural reconstruction of the entire water resources sector within the framework of social and economic reforms. The whole complex of issues includes considerations dealing with the transboundary water allocation in southern Kazakhstan.

### 1.3 Technical level of meliorative systems

Melioration and rehabilitation of irrigated lands have been virtually suspended during recent years. As a result, the engineering and technical level of a considerable part of irrigation systems is such that it hinders their normal functioning.

Reduced funds and deterioration of irrigation systems prevent efficient use of water and land resources and this has led to a large reduction in the crop and fodder production from irrigated lands.

The collector-drainage network is poorly developed on all irrigation systems (it does not exceed 2.14–3.26 linear m/ha). This results in waterlogging and increasing salinity of the soil. Especially difficult is the situation in southern Kazakhstan with the rotation of rice and cotton. Timely planned and continuing maintenance and repairs of the network would lead to increased efficiency and reduced losses of irrigation water, probably saving up to 40 percent of water intake. A reduction of the total water intake for irrigation would make it possible to increase the river flow in the river basin's lower reaches. And this in turn would allow better irrigation of floodplains and stop the processes of degradation of the environment. This is important because free-range livestock grazing plays an important role in the agricultural production of the country. Its further development depends on the availability of good pastures.

### 1.4 Quality of water resources

Irrigation farming is the main source of water pollution. This is due to the use of return water which is discharged back into the river, where it mixes with good quality water, to be re-used for irrigation further downstream. The self-purification property of rivers is reduced, sometimes to a level that endangers aquatic life. The polluted and salinated water may also enter the ground water. When water with increased salt concentration is used for irrigation, this – depending on the salinity of such water – will lead to a decline in crop production. Most irrigation water is used for cultivation of rice and cotton. Return waters from these tracts of land are characterized by a wide range of chemical components and high levels of salts (1.0–5.0 g/litre), ammonium nitrogen (0.6–5.0 mg/litre), pesticides (up to 10 times the maximum allowable concentrations). Some waters in southern Kazakhstan no longer meet water quality standards for domestic and industrial use. This has an impact on health of the people inhabiting the areas so affected.

On average the annual net water volume intake is 20–25 km$^3$ and it is almost fully used. About 75 percent of water intake is used by agriculture, mainly for irrigation. Cultivation of rice and cotton leaves about 25–35 percent as return water to be drained away. This water is enriched in nitrogen resulting from the application of ammonium sulphate, which is removed mainly as nitrate. Its maximum values in water are reached after the application of fertilisers. Nevertheless, water in the major rivers of Kazakhstan is classified as being only slightly polluted.

## 2. FISHERIES IN IRRIGATION SYSTEMS OF KAZAKHSTAN

Kazakhstan is a country with large and diverse water resources. They include northern and central parts of the Caspian Sea, the northern part of the Aral Sea, the large lakes Balkhash and Alakol, hundreds of reservoirs including the largest Bukhtarma (which includes the large lake Zaisan) with an area over 5 000 $km^3$, tens of thousands of medium-size and small lakes and rivers including the Irtysh, Ural, Syr-Darya and Ili.

The availability of such large diversity of waterbodies has stimulated the development of fisheries. When compared with agriculture, fisheries is a relatively young branch of Kazakhstan economy. Its intensive development commenced only in the 1960s. The fisheries targeted the Aral Sea and river fish stocks, and reservoir fish stocks. Fisheries management measures included introductions of new fish species, hatchery production of stocking material and introduction of fish food organisms. New fish processing factories, fish pond farms, lake farms and other enterprises producing stocking material of valuable fish emerged.

All those efforts led to a more than threefold increase in fish production in the country and by 1975 fish catches reached a maximum of 100 000 tonnes (Table 7). It should be emphasized that the valuable sturgeons represented a considerable share in catches. Annual black caviar production reached 885–1 239 tonnes during those years. More recently, as a result of deterioration of ecological conditions on the Caspian Sea, commercial sturgeon fish stocks have dropped dramatically. Today the annual catch is about 250 tonnes.

**Table 7**

Fisheries production (thousand tonnes) in Kazakhstan by Five-Year Plans

| Major fish producing reservoirs | Years of Five-Year Plans | | | | | |
|---|---|---|---|---|---|---|
| | 1975 | 1980 | 1985 | 1990 | 1995 | 2000 |
| The Ural-Caspian basin, total, | 59.48. | 50.38. | 39.3 | 43.8 | 26.7 | 21.6 |
| sturgeon | 2 | 1 | 5.9 | 1.9 | 0.57 | 0.25 |
| The Balkhash-Ili basin | 12.6 | 12.4 | 11.8 | 11.0 | 6.6 | 3.3 |
| Kapshagay reservoir | 0.6 | 0.8 | 1.3 | 1.1 | 0.54 | 0.7 |
| Alakol group of lakes | 4.7 | 3.0 | 2.5 | 2.1 | 1.4 | 0.9 |
| Bukhtarma reservoir | 12.7 | 8.5 | 7.5 | 7.9 | 6.0 | 9.3 |
| Shulbin reservoir | – | – | 0.01 | 0.01 | 0.06 | 0.15 |
| Chardara reservoir | 1.8 | 1.7 | 1.8 | 2.5 | 1.6 | 0.36 |
| Other reservoirs | 8.2 | 9.0 | 8.6 | 0.2 | – | – |
| Total | 100.7 | 85.7 | 72.8 | 68.6 | 42.8 | 36.7 |

Reservoirs play an important role in fisheries. Smaller reservoirs (Table 8) range from several hundred to several thousand hectares and they are used mainly for irrigation, domestic and industrial water supply. Only a small part of the large number of small reservoirs is multipurpose.

Hydroelectric power stations were constructed on five large reservoirs, i.e. Bukhtarma, Chardara, Kapshagay, Shulbin and Ust-Kamenogorsk. Reservoirs located in the north of the country are mainly for power production, water supply for industries, and for fisheries. They are rarely used for irrigation. Reservoirs in the south serve mainly irrigation.

Among the large reservoirs fisheries are well developed on the Bukhtarma in the north, and on Chardara and Bugun in the south.

**Table 8**
Major reservoirs in Kazakhstan

| Reservoir | Volume, $Mm^3$ | | Area, surface (ha) | |
|---|---|---|---|---|
| | At full supply | At minimum supply | At full supply | At minimum supply |
| Bukhtarma | 49 620 | 30 810 | 549 000 | 306 000 |
| Chardara | 5 200 | 4 200 | 90 000 | 28 700 |
| Kapshagay | 28 140 | 6 640 | 84 700 | 51 000 |
| Bugun | 370 | 368 | 68 500 | 4 900 |
| Ters-Aschibulak | 58 | 58 | 24 000 | 2 000 |
| Badam | 61.5 | 59 | 4 800 | 470 |
| Samar | 260 | 215 | 7 500 | 2 200 |
| Topar | 274 | 233 | 4 100 | 3 600 |
| Kengir | 319 | 309 | 3 740 | 1 000 |
| Zhezdin | 76 | 73.6 | 1 760 | 240 |
| Vyacheslav | 410.9 | 377.6 | 6 090 | 980 |
| Sergeev (on the Ishim river) | 693.0 | 635 | 11 680 | 1 920 |
| Seletin | 230 | 200.1 | 3 630 | 210 |
| Upper-Tobol | 816 | 782 | 8 740 | 600 |
| Karatomar | 586 | 562 | 9 400 | 675 |
| Kirov (on the Kushum river) | 63 | 60 | 3 910 | 290 |
| Dungulyuk | 57.4 | 48 | 3 150 | 600 |
| Bitik | 106.7 | 100 | 5 300 | 1 430 |
| Karagalinsk | 280 | 262 | 2 750 | 500 |
| Charsk | 80 | 75 | 1 150 | 1 100 |
| Taschitkul | 620 | 551 | 7 760 | 1 670 |
| Storage lakes of the Irtysh-Karaganda canal | 273 | 210 | 15 600 | 11 500 |
| Shulbin | 263 | 162 | 27 500 | 12 500 |

**Bukhtarma reservoir** was constructed on the Irtysh River in 1960, flooding and enlarging Lake Zaisan. The seasonal drawdown is negligible: in spring (May) within 15–20 days there is a high-water discharge of up to 2.5–3.0 $km^3$ for irrigation of grasslands in the low-lying Pavlodar Oblast. More recently, in connection with the construction and commissioning of the Shulbin hydroelectric power station, the volume of the spring discharge has been reduced considerably. The relatively stable level of Bukhtarma reservoir promotes successful spawning of fish. During the first years of the reservoir the fish stocks mainly consisted of fish of low value, such as roach, ruff, perch and some other species, which represented over 70 percent of the total catch. To enhance the quality of fish stocks, fish species of higher value were introduced, especially during the first several years. Of these zander (*Stizostedion lucioperca*), bream (*Abramis brama*), European cisco (*Coregonus albula*) established themselves and became a regular component in commercial catches. Fish food organisms from lakes Onega and Baikal and from the Caspian Sea were also introduced and now form a valuable component of fish food.

All this, in combination with the new fishing regulations, permitted a twofold increase in fish catches by 1974–1975 as compared with the first years. Fish catches reached a maximum of 12 700 tonnes in the Bukhtarma (Table 7), with a fish yield of 26.0 kg/ha. At the same time the share of poor value fish species decreased to 16 percent while the introduced species made up over 80 percent of the total catch.

Some fish losses are incurred during their passage through turbines of hydroelectric stations because there are no fish protection installations to prevent this. However, this has little impact on fish stocks because in Buktarma reservoir the water intake is more than 60 m below the water surface.

Water pollution, including pesticides, organochlorines, heavy metals and oil products, have an adverse impact on aquatic organisms, but the maximum values do not exceed the maximum allowable concentrations.

**Chardara reservoir** was constructed for irrigation and hydropower production in 1965 on the middle course of the Syr-Darya. Like a majority of irrigation reservoirs in southern Kazakhstan, Chardara is filled in the autumn-winter period (October-March) and drawdown takes place in spring and summer (April-September). Seasonal fluctuations of the water level reach up to 11 m. During the spring and summer the drawdown reduces the reservoir surface to only 15 000– 20 000 ha, which is a quarter or less of the water surface area of the full reservoir. At the same time the water volume is reduced 10–12 times. Such considerable seasonal changes in the volume and the surface area are mainly due to the irrigation water uptake. Water for irrigation enters Kyzylkum Canal, but some water is also used for power generation.

The ichthyofauna of Chardara reservoir formed from the indigenous species of the middle course of the Syr-Darya. From the 33 introduced fish species, 20 species are not present in the reservoir, but 12, including common carp, crucian carp, bream and zander (pike-perch), are of commercial importance. Regular fishing commenced in 1968. During the first eleven years (1968–1978) annual fish catches increased from 400 tonnes to 1 800 tonnes, with an average catch of 1 000 tonnes per year. Bream (52.2 percent) and common carp (22.7 percent) dominated. The largest volume of bream was recorded during 1980–1991, zander occupied second place and common carp third. Annual catches fluctuated from 1 340 to 1 940 tonnes, averaging 1 720 tonnes per year. Commercial fish yield fluctuated from 22 kg/ha to 39 kg/ha, averaging 30 kg/ha.

The most important factors influencing the fish stock formation in Chardara reservoir are hydrological regime which is determined by water uses other than fisheries, hydrochemical regime, and fishery. Reduction of the surface area and shallowing leads to the reduction in the number of spawning areas and to a high mortality of the spawn during the spring breeding period (April 1–May 20). In summer months fish may die due to low concentrations of dissolved oxygen resulting from blue-green algal blooms. Reservoir drawdown also considerably reduces the habitat of benthic organisms and changes their species representation. Benthic organisms are an important food source for some fish species. By autumn, with the drop in water temperature the situation worsens, especially for plankton. There is also a loss of valuable fish species such as zander, roach and bream due to the fry being carried out into the Kyzylkum Canal. In May the maximum number of fry are washed out, with an estimated loss of one million per day. However, the reproductive potential of the fish species still allows an annual fish harvest of 400 to 1 800 tonnes.

Starting in 1992 there was a change in the schedule of the use of the reservoir. The upper part of Chardara reservoir was now located in Uzbekistan. In 1992–2000 high discharges of the Syr-Darya from February to April led to water spilling over the Chardara spillway as well as over the Arnasay emergency spillway. Large quantities of fish fry and fish of all age groups were washed out. The outflow of 800–1 600 $m^3$/sec over the emergency sluice resulted in great fish losses. It is estimated that during those two months, from 1992 to 2000, annual losses of fry of valuable fish species reached 18 to 64 billion. Taking into consideration that the commercial fishing pressure has not changed during the last ten years and the impact of other factors has not changed either, one can assume that the loss of fry and larger fish during floods was responsible for the sharp drop in catches of commercial fish species, from 2 040 tonnes in 1992, to 216 tonnes in 1999.

The following measures are suggested to reduce the rate of fish loss and to increase the stocks of valuable commercial fish species: reconstruction of the Kyzylkum sluice; relocation of the water intake deeper than at present; and construction of a fish protection device on the Arnasay emergency spillway. It is also planned to construct a new reservoir (Koksarai) 120 km distant from Chardara reservoir, to store flood waters which now end in Arnasay.

Since the 1980s the reservoir water has been polluted with pesticides, and this has led to fish mortalities. Over a ten-year period it is estimated that 865 tonnes died because of that. In the more recent years fish mortality has rarely been observed during the pre-spawning and spawning periods as concentrations of pesticides and herbicides have declined as a result of the reduced use of these agrochemicals.

Low value fish species such as silver carp have recently dominated the catches. It is believed that this is because of mass mortalities of common carp during the period 1980–1990, resulting from pesticide poisoning.

The ratio among fish species is continuously changing, with appearance of new species. The recent dramatic reduction in stocks of valuable commercial fish species requires major rehabilitation measures. These include limits on the catch or even closing of the fishery, and intensive stocking of the reservoir with fish species of high value. It has been proposed to intensify stocking of one-year-old common carp, which in four year time should lead to a sustainable carp harvest of 300 tonnes/year. An attempt to close the carp fishery during the period 1997–2000 failed because of lack of enforcement. Low value fish species should be controlled by intensive fishing to reduce the pressure on fish food organisms. This would make them available for fish species of a better value.

Prior to the regulation of the Syr-Darya River the introduced Chinese herbivorous carps were already reproducing in the river. Silver carp entered the Syr-Darya from fish ponds located at Kairakkum reservoir in Uzbekistan. By 1970 the fish were 4–6 year old. In the reservoir they reached commercial quantities for the first time in 1974 when they became a common component of fish catches. To increase their stocks it was recommended to stock reservoir every year with 250 000 two-year-old fish.

Herbivorous fish undertake regular spawning migrations to spawning areas. Recent studies show that 11 000–30 000 (equal to 70–300 tonnes) of silver carp migrate for spawning in the inflowing rivers. Downstream drift of spawn of silver carp takes place in late April, early May. There are no one-year-old fish in the reservoir. This indicates that commercial stocks of silver carp form at the expense of the fish escaping from fish ponds. Over 250 000 fry may have arrived this way, and others drift in from the upstream situated reservoirs and farms on the Syr-Darya in Uzbekistan. According to estimates based on data from trawling surveys, in the year 2000 commercial stocks of silver carp were 74.2 tonnes. As there is a high mortality of spawn drifting into the reservoir it is recommended to collect seed in the Syr-Darya and grow it in ponds to a size suitable for release into the reservoir. This should also increase the reservoir fish production and reduce the pressure on hatcheries which breed stocking material from broodstock kept in fish farms. It is believed that in order to provide a sustainable harvest of 100–120 tonnes, the current stocking rate should be increased 3–4 times.

**Bugun reservoir** is the second largest irrigation reservoir in southern Kazakhstan. It is located in the valley of the Bugun River in the South-Kazakhstan Oblast. It receives water from the Bugun and Arys rivers. The dam is 5 km long, with a height of 17 m, resulting in a maximum reservoir depth of 15 m. The average depth is 5.8 m when full, and 1.4 m at a minimum water level. The reservoir is 13.5 km long and 5.7 km wide. The reservoir fills in autumn and winter and drawdown takes place in spring and summer. Water level fluctuates by up to 10 m. During the three summer months the reservoir loses three quarter to four fifth of the accumulated winter water and its water surface area is reduced to 15 percent of that when full. Bugun reservoir is a classical example of contradiction between the needs of irrigated agriculture and fisheries: irrigation requires drawing water in spring when an increase in water level is in the interest of fishery as fish start to spawn.

During the first years the reservoir fish catches were high and yielded 70 to 80 kg/ha. The total catch was 250 to 360 tonnes/year. Common carp was the major fish species caught, forming 77 percent to 89 percent of the total. Subsequently the fish catches begun to fall rapidly as a result of unfavourable water level fluctuations which had a negative impact on spawning and nursery grounds. The 1977 total catch was 365 tonnes, but in 1979 it was only 101 tonnes. Large fry losses of especially common carp resulted from spilling and pumping. While fish species composition has not changed, there were changes in the species dominance in catches. The present fish stocks in Bugun reservoir are represented by silver carp, crucian carp, eastern bream (*Abramis brama*), zander, asp, wels (*Silurus glanis*), bighead carp, and roach.

By the end of the 1970s fish stocks in Bugun reservoir were very much reduced due to the unfavorable water level during the reproduction period and due to losses of fry pumped into the irrigation canals. The 1982–1984 annual catch ranged from 50 to 96 tonnes, and the yield declined to 14.2–27.4 kg/ha. To enhance the fish stocks, starting in 1985 the reservoir was stocked with fish produced in a hatchery cum fish farm, constructed in the vicinity. The major fish species stocked were common carp and silver carp. As a result the annual catch begun to increase and reached 278 tonnes by 1988. But the release of 650 000 yearlings of common carp during the period 1985–1987 did not make much difference in catches. The high water discharges and the lack of fish protection devices are blamed for an increase in fish losses from the reservoir, with fish entering the major irrigation canal. If such losses are prevented, the stocking would become effective. It is estimated that annual stocking of the reservoir with approximately one million of one- and two-year-old common carp and 200/ha two-year-old silver and bighead carp would result in a 30 percent increase in catches, increasing the annual catch by up to 75 tonnes.

The following two approaches have been identified as the most important for increasing the commercial fish stocks using the existing natural fish feed supply in the reservoir:

- installation of fish protection devices on water discharge into the Turkistan Canal, and
- stocking of the reservoir with two-year-old common, silver and bighead carps.

## 3. DISCUSSION

Fish protection devices were not envisaged under the construction of multipurpose or irrigation dams. This contributed to the disappearance of many valuable fish species initially present in reservoirs, as they were lost in irrigations canals, spilled into depressions, or – in the case of the northern reservoirs – passed through the turbines and spilled. No doubt many species also could not adapt to the sudden change from river to reservoir conditions. The following fish species became very rare or disappeared: Siberian sturgeon (*Acipenser baeri*), nelma (inconnu) (*Stenodus leucichthys*), and ide (*Leuciscus ide*) in Bukhtarma reservoir; ship (*Acipenser nudiventris*) and Balkhash marinka (*Schizothorax argentatus*) in Kapshagay reservoir; ship and Aral barbel (*Barbus brachycephalus*) in reservoir. Installations of fish protection devices is considered to be in the interest of the fisheries.

Thirteen fish hatcheries-cum-fish farms have been constructed and currently operate in Kazakhstan. Most of them are adjacent to reservoirs, including Bukhtarma and Kapshagay and they produce fish for stocking reservoirs. They play a significant role in enhancing stocks of common carp and silver carp. Annually 40 to 50 million fry and fingerlings of such species are released, in addition to some 300–400 million fry of European cisco and peled (*Coregonus peled*).

Small reservoirs in Kazakhstan, ranging from several hundred to several thousand hectares, are used for irrigation, industrial and domestic water supply. They also have extensive fisheries, but largely of a recreational character. Studies of small irrigation reservoirs in southern Kazakhstan (Almaty Oblast) showed that they have good production potential when stocked with one-year-old common carp, grass carp and silver carp in polyculture. Better utilisation of small reservoirs is at present hampered by shortage of stocking material.

# THE USE OF IRRIGATION SYSTEMS FOR SUSTAINABLE FISH PRODUCTION: UZBEKISTAN

by
**Bakhtiyar Kamilov**
**Institute of Aquaculture, 21a, Chilanzar-10, Tashkent, 700123, Uzbekistan**

## ABSTRACT

In Uzbekistan capture fisheries is practised in freshwater irrigation and multiuse reservoirs, and in lakes for residual water storage. Two groups of lakes are of major importance for capture fisheries: the Amu-Darya Delta, where there are 20 lakes varying from 4 000 to 15 000 ha in an area covering a total of 97 000 ha, provides about 1 500 tonnes of fish annually, and the Aydar-Arnasay lake system, where situated off the middle course of the Syr-Darya River.

In 2000, 1 600 tonnes of fish were captured. Before independence, fish farms annually stocked up to 15 million one-year-old common carp, bighead carp, silver carp and grass carp in reservoirs and lakes with capture fisheries. In the Aydar-Arnasay lake system the fish yields without stocking during the 1970s and 1980s reached a maximum of 15 kg/ha, and after stocking they increased to a maximum of 25 kg/ha (Kamilov *et al*., 1994). In the 1990s the stocking rate was greatly reduced because of budgetary problems. It is believed that future strategies of fishery development in irrigation waterbodies, apart from continuing to pursue the stocking of reservoirs, should focus on small-scale aquaculture in reservoirs, canals and lakes, with strong involvement of villagers and local administrators.

## 1. INTRODUCTION

Central Asia occupies an area of about 2 million km$^2$ situated deep in Eurasia. This region is enclosed from the south and east by mountains and open to the north and west. About 70 percent of Central Asia is covered by steppes and deserts and 30 percent by mountains. Five countries form this region: Kyrgyzstan (198 500 km$^2$), Tajikistan (143 100 km$^2$), Turkmenistan (448 100 km$^2$), Uzbekistan (447 000 km$^2$) and the southern part of Kazakhstan. The basin of the Aral Sea occupies the central part. The northwestern part of Uzbekistan is represented by the Turan plain covering more than 35 500 km$^2$, the eastern part is covered by the Tien Shan mountain system, which represents about 90 600 km$^2$ or almost 20 percent of the country's area. Uzbekistan extends from an elevation of 10 m in the Sarykamysh depression to 4 640 m in the western Tien Shan (Anon., 1993).

Due to the landlocked location of the region and it being open to the north, the climate is extremely continental, with high aridity. Large seasonal and daily fluctuations in temperature are characteristic. Summer is dry, cloudless and very hot; in July, the warmest month, the average monthly temperature is 25–31 °C; from June to the first half of August the maximum temperature usually reaches 40 °C and more. Only in the mountains is the temperature lower. The dry cold Siberian air mass has free access from the north where there are no mountains. The average January air temperature varies from –8 to –12 °C in the northwest, –6 ° to –8 °C in the mountains, and 0 to +2 °C in the south (Anon., 1993).

Central Asia is an arid region with low precipitation: about 20 percent receives less than 100 mm per year, 90 percent less than 300 mm. Ninety-five percent of the precipitation is generated from moisture entering the region from outside. Usually 30–50 percent of the total rainfall is in spring, 25–40 percent in winter, 10–20 percent in autumn and 1–6 percent in summer. Rivers, lakes and ponds usually freeze from late December – early January to mid-February. There are many hours of sunshine: 2 500–3 000 h per year, and the total annual insolation is between 140 and 160 kcal per $km^2$.

Due to the climate, deserts and arid steppes cover the plains of Central Asia. Large rivers such as the Amu-Darya and Syr-Darya can only exist because of the presence of high mountains, from which they are fed predominantly snow- and icemelt water (Shultc, 1966). Both rivers enter the Aral Sea and their catchments constitute a major part of the Aral Sea catchment. The Aral Sea has no outflow. Uzbekistan is located in the basin of the Aral Sea between the Amu-Darya and Syr-Darya. Its lowlands are dominated by the rivers Amu-Darya, Syr-Darya, Zarafshan and Kashka-Darya. Zarafshan and Kashka-Darya, which were in the past tributaries of the Amu-Darya, do not enter it any more because they have been diverted for irrigation and what flow is left dries up in the desert.

The rivers receive about 90–95 $km^3$ of water largely from the mountains located in Tajikistan and Kyrgyzstan, with only 10 $km^3$ forming in Uzbekistan mountains. Water consumption in Uzbekistan has stabilized at about 62–65 $km^3$ per year. About 85 percent (53–55 $km^3$) of the water is used in agriculture, 12 percent (6 $km^3$) in industry, and 3 percent (1.7 $km^3$) as municipal supply. Agriculture is therefore the main water consumer.

## 2. IRRIGATION SYSTEMS

Although the basin of the Aral Sea is mostly covered by deserts, since ancient times it has been known as a region with highly developed agriculture. This has been possible only through the development of irrigation. For the last 2000 years, irrigated agriculture has expanded far from rivers, such as in the Fergana Valley and Tashkent, and even deep into deserts to form oases, such as Khorezm and Bukhara. Without irrigation those places would still be deserts.

In the second half of the twentieth century Uzbekistan was the main cotton producer and even today it is the fourtn largest producer of cotton in the world, with 20 percent of the total world production. But the cotton monoculture causes ecological problems. While rice and wheat are also cultivated, cotton is still the main culture. In the basin of the Aral Sea the irrigated area during 1925–1980 increased from 2.0 to 7.2 million ha. For one hectare of irrigated land one uses from 16 000 to 18 000 $m^3$ of water, which is about two times more than in other parts of the world cultivating the same crops. A huge and extensive network of irrigation and drainage canals has been created. About 73 percent of the irrigated land is set aside for cotton production, which is unprecedented in the world's agricultural practice. Under the centrally planned economy, the main idea was to use extensive irrigation; however, canals constructed for this purpose are not efficient and water losses from such irrigation networks are estimated at about 40 $km^3$ annually. This amount of water would be enough to stabilize the Aral Sea at its current level.

In Uzbekistan cotton, rice and wheat yields average only one third of those elsewhere. Uzbekistan uses mainly furrow and flood irrigation. Such irrigation causes increased soil salinity that requires more water for soil washing. In cotton monoculture this leads to soil degradation.

Under the impact of man, the river network of the Aral Sea basin has changed. Not only do the waters of Zarafshan and Kashka-Darya not reach the Amu-Darya anymore, but water taken for irrigation from the middle course of rivers, if not too saline after being used, is returned as drainage water to the river downstream. If the salt concentration is too high, the water may be discharged into lakes, some of which have been formed entirely from such water (Nikolsky, 1938; Shultc, 1966). Since the 1960s irrigation and drainage systems tie together different river basins into one network.

Irrigation systems in Uzbekistan include reservoirs, irrigation canals, drainage canals, and lakes for residual water. Before the end of spring, water is collected in reservoirs and in summer it is drawn off until autumn. Water taken for irrigation seeps into soil and enters groundwater. Groundwater may re-enter the river and also lead to formation of swamps. While entering the river downstream this drainage/groundwater boosts up river discharge, but it also changes water quality.

Before the first half of the twentieth century water uptake for irrigation did not upset the water balance of the Aral Sea, but since the 1960s the flow redistribution has caused irreparable damage to the whole ecosystem. Irrigation has changed the water regime in the whole of the Aral Sea basin.

## 3. ARAL SEA CRISIS

Uzbekistan shares the Aral Sea, one of the largest lakes in the world, with Kazakhstan. In 1960, the Aral Sea had an area of about 68 000 km$^2$ and a volume of 1 061 km$^3$ (Micklin, 1988). Before 1960, the Syr-Darya and Amu-Darya rivers annually discharged into the Aral Sea about 56 km$^3$ of water, and a further 8 km$^3$ came in the form of precipitation, and as groundwater flow. The mean annual evaporation from the sea surface was 63 km$^3$. The water level of the Aral Sea was about 53 m above sea level (a.s.l.). As a result of the intensive uptake of water for irrigation up until the 1990s, the annual water runoff reaching the Amu-Darya and Syr-Darya river deltas was reduced to 5 km$^3$, and in some years the rivers virtually stopped flowing into the sea. By 1992 the Aral Sea water level had dropped to 37 m a.s.l., the surface area was reduced to 34 100 km$^2$, salinity reached 34–37 g/litre as compared with 9–10 g/litre in the 1960s. Today the sea coast is 60–80 km from the original coastline.

Waterbodies in the basin of the Aral Sea can be grouped as follows:

- natural waterbodies (rivers and streams, lakes);
- artificial freshwater waterbodies (irrigation canals, reservoirs, ponds);
- artificial saline waterbodies (drainage canals, lakes for residual water storage).

In Uzbekistan there are more than 600 large and small rivers. Only a few of them, those in the mountains, are not affected by irrigation. Only the large rivers Amu-Darya, Kashka-Darya, Zarafshan, Syr-Darya and their tributaries have fishery importance. The Amu-Darya, 1 440 km long, has the highest discharge of about 78 km$^3$. The Syr-Darya has a discharge of 36 km$^3$ and is 2 140 km long. During the last decades, all natural lakes have been impacted by large-scale irrigation development. Some lakes have dried up, others have been used for residual water storage. In the middle and lower courses of Uzbekistan rivers there are practically no natural lakes left whose water regime would not be affected by salinity and by the irregular discharges of drainage water.

In the countries of Central Asia there are about 60 reservoirs with a total volume of 61.6 km³. They were constructed in the basins of all large rivers. In the basins of the two major rivers, Amu-Darya and Syr-Darya, there are 39 reservoirs (Tables 1 and 2). The total water surface of reservoirs with fisheries importance is 3 310 km² (Nikitin, 1991). Some of the large Uzbekistan reservoirs are important for fisheries, such as Tudakul, Shorkul and Mezhdurechye.

**Table 1**

Reservoirs in the Aral Sea basin

| River basin | Number of reservoirs | Area km$^2$ | Volume km$^3$ |
|---|---|---|---|
| Syr-Darya | 22 | 850 | 34.5 |
| Amu-Darya | 17 | 1 460 | 23.3 |

**Table 2**

Reservoir distribution by size

| Volume, million m$^3$ | Amu-Darya basin | Syr-Darya basin |
|---|---|---|
| | Number of reservoirs | |
| 1–50 | 5 | 4 |
| 50–500 | 6 | 13 |
| >500 | 6 | 5 |
| Total | 17 | 22 |
| | Volume of reservoirs, million m$^3$ | |
| 1–50 | 110 | 112 |
| 50–500 | 1 490 | 1 543 |
| >500 | 21 700 | 32 850 |
| Total | 23 300 | 34 505 |

The system of irrigation canals is well developed and has a total length of about 150 000 km. Only 5–6 large main canals, with a length of 100–350 km and a capacity of 100–300 m³/sec each, are at present of fishery significance. These are the Yuzhnogolodnostepsky main canal, Karshi main canal, Amu-Bukhara main canal and several others. In most canals water flows by gravitation. The Karshi main canal and Amu-Bukhara main canal use pumping .

There are about 100 000 km of drainage canals in Uzbekistan. For fisheries only the large main collectors with more than 100 km length and water flow rates of 40–100 m³/sec each are important. The annual discharge of some of these collectors is comparable with that of some rivers, e.g. Ozerny (1.2 $km^3$) and Centralno-Golodnostepsky (1.5 $km^3$).

Lakes used for or artificially created for residual water storage are important for fisheries. Those of importance for fisheries cover about 7 000 $km^2$, a surface area of about twice that of all reservoirs. Most of the lakes function for many years. They do not experience major seasonal changes. After the demize of fisheries in the Aral Sea, the Aydar-Arnasay lake system and the lakes of the Amu-Darya delta are the major waterbodies in this category supporting fishery in Uzbekistan. Due to the current problem of harmonizing the use of the Syr-Darya among the riparian countries, the Aydar-Arnasay system is now receiving large volumes of water, and as a result of that it now covers more than 4 000 $km^2$ which makes it the largest artificial lake in the region.

In the 1960s, the delta of the Amu-Darya had some 40 lakes with a total water surface of about 100 000 ha; now there are only some 20 lakes, but they have a total water surface of approximately 115 000 ha. It is a direct result of the restoration of the main lakes and appearance of new isolated ones on the dried Aral seabed. These waterbodies are maintained almost completely with collector-drainage waters. Furthermore, along the Amu-Darya many large lakes with saline water formed, including Sarykamysh (330 000 ha) and Dengizkul (26 000 ha).

## 4. FISH FAUNA

Prior to large-scale irrigation efforts the indigenous fish fauna in the Aral Sea catchment rivers and lakes was little affected by human activities. Kamilov and Urchinov (1995) listed for Uzbekistan 84 species of fish, including those which were rare and those which were introduced. The ichthyofauna has undergone major changes as a result of water regulation and introductions of fish species from outside the Aral Sea basin (Kamilov, 1973; Kamilov *et al.*, 1994). Some species disappeared or became rare, such as three species of endemic shovelnoses (*Pseudoscaphirhynchus kaufmanni*, *P. hermani*, *P. fedschenkoi*), ostroluchka (*Capoetobrama kuschakewitschi*), minnows (*Alburnoides bipunctatus, A. taeniatus, A. oblongus*) and Zarafshan dace (*Leuciscus lehmanni*), because they have been unable to adapt to the new environment, or because dams blocked their spawning migrations (spiny sturgeon *Acipenser nudiventris*, Aral barbel *Barbus brachycephalus*). Some species such as gudgeons (*Neogobius fluviatilis, N. melanostomus, Pomatoschistus caucasicus, Proterorhinus marmoratus*) and Baltic herring (*Clupea harengus membras*), introduced in the Aral Sea, became established for a while, but later on disappeared as a result of increasing salinity and other changes in the Aral Sea environment.

During 1960–1990 a number of fish species from outside the region were introduced in a number of irrigation waterbodies of Central Asia. Pikeperch and bream were released into reservoirs and lakes of the rivers Zarafshan, Kashka-Darya and the middle courses of the Syr-Darya and Amu-Darya. Silver carp (*Hypophthalmichthys molitrix*), grass carp (*Ctenopharyngodon idella*), bighead carp (*Aristichthys nobilis)* and snakehead (*Channa argus warpachowskii*), introduced from the Far East, were stocked in fish farms in the Tashkent area and from there the hatchery-produced stocking material was regularly stocked into lakes and reservoirs. Three species of buffalo (*Ictiobus cyprinellus*, *I. bubalus*, *I. niger*) and channel catfish (*Ictalurus punctatus*) were also introduced into fish farms but they did not enter rivers except the last species which entered the Syr-Darya. Rainbow trout (*Oncorhynchus mykiss*),

Sevan trout (*Salmo ischchan issykogegarkuni*), peled (*Coregonus peled*) and lake herring (*Coregonus sardinella*) were released into Charvak reservoir in the Tashkent area where they are now established.

Many species spread throughout the basin via the connecting major canals. Some species started to breed in both the irrigation and drainage canals. Fish stocks in canals were not managed. In the 1970s–1980s management concentrated on stocking fingerlings and one-year-old marketable fish, the stocking material of which was produced in fish farms. Silver carp, grass carp, common carp and bighead carp were regularly stocked in reservoirs and lakes for residual waters storage. This resulted in fish yields increasing by 5–15 kg/ha. After 1991 stocking continued only in the Aydar-Arnasay lake system and several other large waterbodies.

## 5. FISHERIES

Until 1960 fishing was concentrated on the inshore waters of the Aral Sea and the deltas of the inflowing major rivers. In the 1960s, Uzbekistan fishery concentrated only on the Aral Sea, with an average annual catch of 25 000 tonnes. At that time only one small fish farm and one hatchery existed near Tashkent. Fisheries were government-owned, but several fisheries cooperatives also operated on the Aral Sea. The major fish species captured were common carp (*Cyprinus carpio*), bream (*Abramis brama*), barbel (*Barbus brachycephalus*), roach (*Rutilus rutilus*) and shemaya (*Chalcalburnus chalcoides aralensis*). Less common were wels (*Silurus glanis*), pike (*Esox lucius*), asp (*Aspius aspius*), sturgeon (*Acipenser nudiventris*), pikeperch (*Stizostedion lucioperca*). During the 1960s–1970s, fish catches decreased sharply. In 1983, the last year of the Aral Sea fishery, only 53 tonnes were caught.

Fisheries in Uzbekistan had to find new sources of fresh fish. During the 1970s fishing fleets were transferred from the Aral Sea to Lake Sarykamysh and the Aydar-Arnasay lake system. During this period up to 6 000 tonnes of fish were caught annually in these lakes. While in 1964 the catch in Aydar-Arnasay lakes was only 26 tonnes, in 1971 it was 512 tonnes and a maximum of 4 200 tonnes was captured in 1988, corresponding to a yield of approximately 25 kg/ha.

But capture fisheries under conditions of irrigation could not replace the quantity of fish lost from the Aral Sea. The Ministry of Fisheries had to do as best as possible to develop fisheries in the new waterbodies, whatever the constraints to fish production arising from their management for irrigated agriculture. The Government of Uzbekistan and the former All-union Ministry of Fisheries developed a large-scale development programme of pond fish culture and fisheries in inland waterbodies. That programme included creation of new fish farms and fishing enterprizes in all regions of Uzbekistan, testing and implementation of new technologies, establishment of research centres, specialist training, etc.

Dissolved salt concentrations in rivers are from 0.5–2.5 g/litre, in reservoirs 0.5–2.5 g/litre, in lakes for residual waters storage 4–18 g/litre. Freshwater of less than 1 g/litre is only in the upper catchments of the rivers and tributary streams and in the upper parts of the middle courses of major rivers. Further downstream all rivers and associated lakes receive return waters. From 1960 to 1995 water salinity in the lower Amu-Darya increased from 0.5 to 1.2 g/litre. Today water salinity near the Termez city, which is situated between the upper and middle course of the Amu-Darya, is about 0.3–0.5 g/litre. Further downstream it raises to

1.2 g/litre. During the last decades water salinities in rivers ranged from 0.5 to 2.0 g/litre, in reservoirs from 0.5 to 2.5 g/litre, in lakes formed from residual waters and in natural lakes receiving such waters were from 3 to 20 g/litre.

While irrigation reservoirs have good water quality, they also have some limitations, such as unseasonal water level drawdown which conflicts with fish reproduction. Also there are no structures which would prevent all developmental stages of fish from entering irrigation systems, in which they perish. Thus, in the lower Amu-Darya up to 90 percent of larvae and fry enter canals and end on irrigated fields (Pavlovskaya, 1995). On the other hand, large connecting canals may be beneficial for fish distribution. For example larvae and fry of the middle course of the Amu-Darya migrate through the Amu-Bukhara Main Canal into Tudakul reservoir where they significantly contribute to maintaining fish stocks in this reservoir.

By the end of the 1980s, the annual volume of drainage water in the basin of the Aral Sea reached about 33 km$^3$ which was about 60 percent of the total river discharge into the Aral Sea. This volume included 17 km$^3$ in the basin of the Amu-Darya, 13 km$^3$ in the Syr-Darya and 3 km$^3$ in the Zarafshan and Kashka-Darya. About 20 percent of the 33 km$^3$ was returned into rivers and 10–13 km$^3$ was diverted into depressions where this water eventually created Lake Sarykamysh (3 000 km$^2$), the Aidar-Arnasay lake system (at present more then 4 000 km$^2$), and a number of smaller lakes.

For a while lakes for residual water storage were more preferred for capture fishery than reservoirs as they behaved like lakes, i.e. their water level was not affected by drawdowns. But, after the dissolution of the USSR the regional system of water resource management in the Aral Sea basin became fragmented as a result of each country obtaining full independence. At present water resources in the basin of the Aral Sea are regulated from five centres, one in each country of Central Asia. This has already caused a number of problems. For example, during the period 1991–2001 huge amounts of Syr-Darya water had to be discharged into the Aydar-Arnasay lake system, which has no outflow, therefore the water cannot be reused.

As all good land was allocated for agriculture fish farms had to be established on land with poor soil. In spite of that, with good management they all functioned well between 1970 and 1990. In the 1980s the total annual fish production in Uzbekistan averaged 30 000 tonnes per year. From this, 7 000 tonnes came from capture fisheries and 23 000 tonnes from fish pond culture (Table 3). Besides the capture and culture, the fisheries organisations also dealt with fish transport, storage and marketing of about 60–70 000 tonnes of marine fish imported from other regions of the former USSR. This comprehensive fishery programme was possible under the very centralized system, where the fisheries were under the USSR Ministry of Fisheries, which was assisted by the appropriate Uzbekistan authorities.

After the disintegration of the USSR, under conditions of economy reforms towards the free market economy the Government of Uzbekistan decided to privatise the fisheries. Starting in 1994, it discontinued its financing. Fishermen found themselves in the new unfamiliar conditions of a market economy. The overall economic crisis and the loss of economic links with producers of equipment in the former USSR have also adversely affected fisheries. Over the last ten years the fishing equipment has much deteriorated. The number of fishing boats, set nets and seines dropped. In the 1990s there were only 20 fishing boats with 130 horsepower engines, 40 boats with 20 to 60 horsepower engines, and 250 other types of motorized boats. All fishery companies together had only 5 000 gillnets and 36 beach seines, which are now worn out. Table 3 gives information on reservoir, lake and river capture fisheries for selected years. After a major decline in catches, which reached the lowest value in 1996, there has been a slow recovery.

The main limitation in aquaculture was the absence of formulated fish feed. In 1995, the total fish production was 5 600 tonnes. Table 3 gives data on aquaculture production in Uzbekistan prior to and after gaining independence. The same trend can be seen as for the capture fisheries with the lowest production reached in 1996 (Table 4), this to be followed by a slow recovery. The 20 existing fish farms in Uzbekistan were established along the irrigation network: twelve of them have freshwater, eight contain drainage water with a salinity of 5–6 g/litre.

**Table 3**

Capture fisheries and aquaculture in Uzbekistan (in thousand tonnes)

| Year | Capture fisheries | | | Aquaculture | Total |
|---|---|---|---|---|---|
| | Lakes* | Reservoirs | Rivers | | |
| 1980s** | 5.5 | 1.0 | 0.5 | 23.0 | 30.0 |
| 1994 | 2.0 | 0.8 | 0.3 | 14.6 | 17.7 |
| 1996 | 1.2 | 0.3 | 0 | 5.0 | 6.5 |
| 1999 | 3.1 | 0.4 | 0 | 5.6 | 9.1 |
| 2000 | 2.7 | 0.3 | 0 | 6.2 | 9.2 |

* – lakes used for residual water storage
** – average for a decade

**Table 4**

Aquaculture production (thousand tonnes) in Uzbekistan by type of water

| Year | Fish farms with freshwater | Fish farms with saline water | Total |
|---|---|---|---|
| 1990 | 13.2 | 7.6 | 20.8 |
| 1993 | 12.8 | 6.1 | 18.9 |
| 1996 | 3.8 | 1.2 | 5.0 |
| 1999 | 4.1 | 1.5 | 5.6 |

Due to lack of access to oceans, as well as the demise of the Aral Sea fisheries, the future of Uzbekistan's fisheries lies in aquaculture and enhanced capture fisheries. Extensive fish pond aquaculture is the most important sector of the fisheries industry, providing 60 percent of today's total fish production. About 20 companies own hatcheries, which induce-breed and farm fish, mainly silver carp, bighead carp, and grass carp, with common carp reproduced both artificially and naturally. The fish are grown on farms to market size. In the 1990s the total area of ponds reached 10 400 ha, with sizes of the individual ponds ranging from 10 to 150 ha.

All waterbodies of irrigation systems (with the exception of two reservoirs serving hydropower production) and the fish in them are still state-owned and managed by the Ministry of Agriculture and Water Management. Fisheries are private and any interested party can develop them. But most enterprises have joined in the "Uzriba" Corporation that was established from the former Fisheries State Committee of Uzbekistan. "Uzriba" consists of 20 fishing and aquaculture production, aquaculture engineering, research and fish disease control organizations.

## 6. CONCLUSIONS

The river systems in the Aral Sea basin have been heavily modified for irrigation. The network of reservoirs and canals was created under the conditions of the centrally planned economy of the former USSR, when the priority of agriculture was cotton production. Fisheries was given a low priority, nevertheless there was allocation in the budget for its development, especially in the irrigation reservoirs. The large-scale diversion of water for irrigation and damming of rivers led to changes in the fish species composition in the modified rivers and in reservoirs. As irrigation demand consumed most of the available water, water discharges into the Aral Sea greatly diminished, the sea started desiccating and its water salinity increasing. This led to the loss of fish stocks and the Aral Sea fisheries ended in 1983.

A comprehensive programme was initiated which led to the establishment of new fisheries in reservoirs. One of the reasons for that was to develop an alternative fish supply to that which was lost from the Aral fisheries. During the 1960s and 1980s, 20 fishing companies were created to manage the fisheries in all large lakes, reservoirs, and in lakes for residual water storage. All capture fishery and aquaculture companies were state-owned and financed from the government budget. Special attention was paid to the production in fish farms of material for regular stocking of reservoirs as well as for production of market size fish. In the 1980s up to 30 000 tonnes of fish were annually produced in Uzbekistan, all fisheries using water of the irrigation system. During those years there was a close cooperation between the All-union (i.e. USSR) Ministry of Fisheries and the government of the Republic of Uzbekistan.

Dissolution of the USSR, problems during the period of economic reforms, privatization and cutting off the financial support of the government led to substantial decrease in fish catches and aquaculture production, with the production falling to as low as 5 600 tonnes per year. This was followed by a partial recovery by the end of the 1990s as private farmers started to be better acquainted with the new economy. In 2000, a total of 9 000 tonnes of fish were captured and produced in aquaculture.

Uzbekistan has a great potential for aquaculture and capture fishery development through enhancement, also because the climatic conditions are very suitable. But there is a need for a change in attitude. Fisheries has never been on the list of government priorities and today support by the government is minimal. The privatized Uzbekistan fisheries industry, united in the corporation "Uzbalik", has as an objective the rehabilitation of fisheries in larger waterbodies of Uzbekistan. But there is a need to develop aquaculture, especially on a small-scale. Small-scale aquaculture could be developed in reservoirs, canals and lakes, with participation of villagers and local administration. At present the Government deals only with "Uzbalik".

Today small-scale development of aquaculture is also limited by the lack of funds. Farmers do not yet understand the mechanism of using bank credits for farming, and this includes also fish farming. At present there is no competition among producers, because local administration traditionally and preferentially protects local fish farms. Once the credit mechanism is fully understood, with the help of the local administration the new small-scale aquaculture ventures should be easy to implement.

From the technical point of view there is a great need for the development of model small-scale aquaculture farms which could be established on irrigation systems in the countries of Central Asia. The models should be based on pilot studies of how to achieve the following production:

- extensive culture of carps for average production of 1.5–2.0 tonnes/ha;
- semi-intensive culture of carps with feeding (4–6 tonnes/ha);
- culture of high value fish species (catfish, sturgeons, snakehead);
- experimental culture testing other fish species of high value.

To carry out such studies successfully will require international assistance, both advisory and financial.

## REFERENCES

Anon., 1993. Environment protection and use of natural resources in the Republic of Uzbekistan. Report of the State Nature Protection Committee of Uzbekistan, Tashkent. 90 p. (In Russian)

Kamilov, G.K. 1973. Fishes in Uzbekistan reservoirs. Fan, Tashkent. 234 p. (In Russian)

Kamilov, G.K., Karimov, B., Khakberdiev, B., Salikhov, T., Kamilov, B., Pridatkina, N.V., Abdullaeva, L., Taraskin, A., Al Madjid, Yuldashev, M., Nazarov, M., Gerasimova, O. 1994. Uzbekistan waterbodies and their importance for fisheries. Two volumes. Tashkent State University, Tashkent. (In Russian)

Kamilov, G. &. Urchinov, Zh.U. 1995. Fish and fisheries in Uzbekistan under the impact of irrigated agriculture. FAO Fisheries Circular No. 894: 10–41. FAO, Rome.

Micklin, P. 1988. Desiccation of the Aral Sea: A water management disaster in the Soviet Union. Science 241, pp. 1170–1176.

Nikitin, A.M. 1991. Reservoirs of Central Asia. Gidrometeoizdat, Leningrad 165 p. (In Russian)

Nikolsky, G.V. 1938. Fishes of Tajikistan. Academy of Science of the USSR, Moscow-Leningrad, 228 p. (In Russian)

Pavlovskaya, L.P. 1995. Fishery in the lower Amu-Darya under the impact of irrigated agriculture. FAO Fisheries Circular No. 894: 42–57. FAO, Rome.

Shultc, V.L. 1966. Rivers of Central Asia. Gidrometeoizdat, Leningrad. 691 p. (In Russian)

# UZBEKISTAN IRRIGATION SYSTEMS AND THEIR MANAGEMENT POTENTIAL FOR FISHERIES IN REGIONAL CONTEXT

by

**P. Umarov**
**11 Kazaksu-4, Tashkent, 700187, Uzbekistan**

## ABSTRACT

Irrigation in Uzbekistan has been practised for thousands of years, but the large-scale infrastructure was constructed only during the Soviet period in response to the growing demand for cotton. Eventually cotton production used more than 90 percent of total water resources available in the country.

Since independence in 1990, Uzbekistan, as the major water user in Central Asia, has found itself in a difficult situation as a result of declining water management and growing water deficit, as well as lack of funds for further irrigation development. Water management activities have lately focused on operation of old irrigation systems and negotiating transboundary water problems. Fisheries, introduced in the region several decades ago in order to compensate for the loss of the Aral Sea fish catch, use irrigation and drainage systems. Under the favourable climatic conditions of Uzbekistan these waterbodies have a high potential for fish production.

However, constraints arising from the existing problems with water management of irrigation systems as well as those caused by other uses of water in Uzbekistan negatively affect fish production. Over the last ten years the fish production from waterbodies of irrigation systems decreased threefold. All fisheries are now privatized, but the privatization has resulted in decline of the formerly well developed fisheries research and management structures. Coordination among fisheries institutions responsible for maintaining sustainable fish yields in a great diversity of waterbodies serving and arising from irrigated agriculture has been disrupted. Linkages between the irrigation and fishery institutions have been lost, and there is lack of new initiatives and research projects facilitating and promoting the use of irrigation systems for fish production.

There is an urgent need for improving intersectoral cooperation in integrated water management, as well as a need for pilot research projects which would test the best fish enhancement practices for Central Asia. While the use of water for irrigated agriculture is still government priority, and the proposed new reforms in agriculture and water management, when implemented, will create a favourable environment for fish production, Uzbekistan needs international support, including funding. Further assistance is needed to establish a regional information network for exchanging information on the best possible use of irrigation systems for fish production. This also requires the international assistance of specialized fisheries organizations and institutions, such as FAO, NACA, etc. Introducing the best existing practices of fish production in irrigation and drainage waterbodies, selected from global experience, is of high importance for improving food security in the Region. While at present no regional network exists to deal specifically with the use of irrigation systems for fish production in arid countries, there is a potential for cooperation in the use of irrigation systems for fish production at the regional level. It is believed that the Interstate Commission for Water Coordination (ICWC), based in Tashkent, Uzbekistan, is a body capable of taking up this function.

# 1. STATUS OF THE USE OF IRRIGATION SYSTEMS FOR FISH PRODUCTION IN UZBEKISTAN

## 1.1 Status of irrigation development

Water is one of the major factors on which life and development in Central Asia depend. Irrigation was one of the main uses of water in the region from time immemorial, its origins dating back to the seventh millennium B.C. By the beginning of the twentieth century there were 2.5–3.5 million ha of irrigated lands in the Region, supported by an extensive irrigation network. When Uzbekistan was part of the Soviet Union water resources in Uzbekistan were subject to large–scale engineering modification, as demanded by the growing irrigated agriculture, basically for cotton production.

Water resources in Uzbekistan are part of the water resources of the Aral Sea basin (Map 1). This basin includes two major rivers of Central Asia: the Amu-Darya and the Syr-Darya, which are the main sources of surface flow. The portion of water originating directly from the territory of Uzbekistan is 6.3 percent in the Amu-Darya basin and 16.5 percent in the Syr-Darya basin, which represents 9.6 percent of their total flow (Anon., 1997) (Table 1). The water is stored in numerous reservoirs (Table 2 ) from where it is distributed on demand.

**Table 1**

River discharges in the Aral Sea basin (km$^3$/year)

| State | River basin | | Aral Sea basin | |
|---|---|---|---|---|
| | Syr-Darya | Amu-Darya | km$^3$ | percent |
| Kazakhstan | 2 624 | – | 2 626 | 2,1 |
| Kyrgyzstan | 27 605 | 1 604 | 29 209 | 25.1 |
| Tajikistan | 1 005 | 59 578 | 60 583 | 52.0 |
| Turkmenistan | – | 1 549 | 1 549 | 1.2 |
| Uzbekistan | 6 167 | 5 056 | 11 223 | 9.6 |
| Afghanistan and Iran | – | 11 593 | 11 593 | 10.0 |
| Total Aral Sea basin | 37 203 | 79 280 | 116 483 | 100 |

The use of water resources by different sectors of Uzbekistan is shown in Table 3. The main water use beneficiaries in Uzbekistan are irrigated agriculture, which uses 90 percent of the available water, drinking and rural water supply, industries and fisheries. These sectors also generate 28.2 km$^3$ of return waters (Anon., 2000). Intensive development of irrigation and drainage in the Aral Sea basin has had two major impacts on water quantity and quality in the rivers: a major freshwater uptake for irrigation, and generation of polluted return water of elevated salinity.

More then 50 percent of the total irrigated area in Central Asia (4.3 million ha) is located in Uzbekistan. While the area under irrigation during the 10 years of independence remained the same there were some changes in the crop pattern. Cotton still remains the priority crop; however, its share in irrigated agriculture has decreased from 50 to 40 percent. The share of cereals (wheat, rice, maize etc.) has increased from 13 to 30 percent. The share of fodder crops has remained the same at 20 percent of the total (Anon., 2001).

**Map 1**

Location of the Aral Sea basin

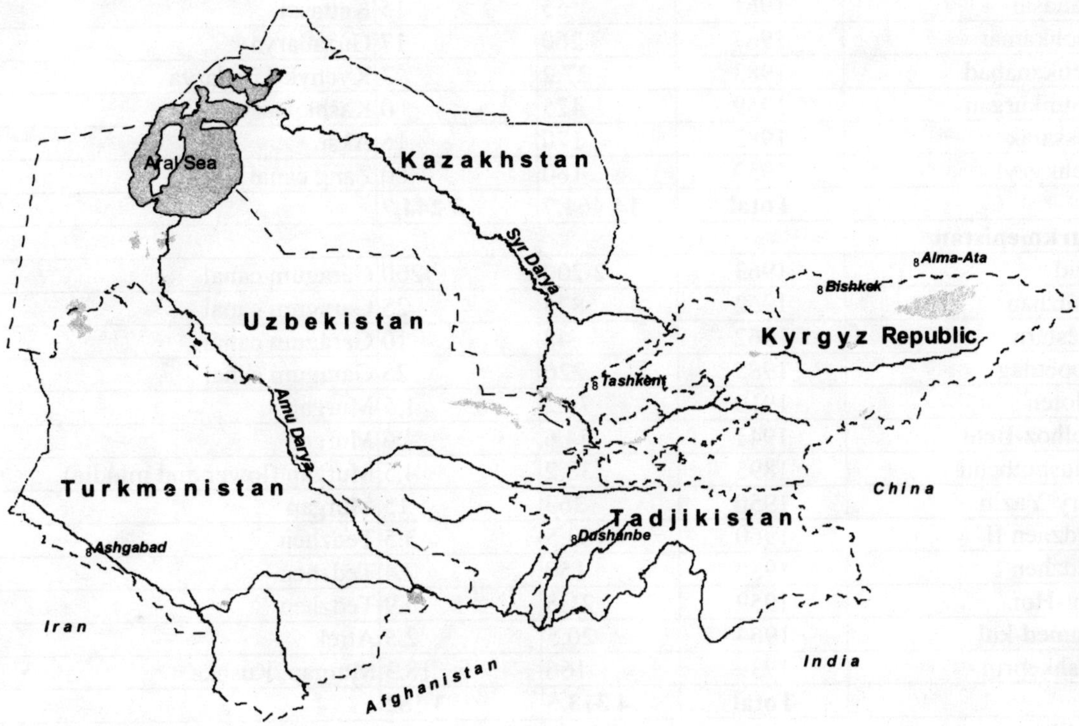


The irrigation infrastructure of Uzbekistan comprises an interconnected irrigation system of canals and drainage collectors, with freshwater and drainage (return) water flows. In Uzbekistan there are 28 000 km of main and inter-farm irrigation canals and 168 000 km of in–farm irrigation canals. The total length of main and inter-farm collectors is more than 30 000 km and there are 107 000 km of in–farm collectors (Anon., 2000) (Table 4).

**Table 2**

Reservoirs in the Aral Sea basin (SIC ICWC data base) (million $m^3 = Mm^3$)

| Name | In exploitation since | Total capacity $Mm^3$ | Dead capacity $Mm^3$ | Source |
|---|---|---|---|---|
| **Amu-Darya River basin** | | | | |
| **Uzbekistan** | | | | |
| Tuyamuyun | 1980 | 7 800 | 2 550 | Amu-Darya |
| Tudakul | 1986 | 1 200 | 50 | Andijan Great canal |
| Talimazhan | 1978 | 1 525 | 125 | Karshi main canal |
| South Surhan | 1962 | 800 | 210 | Surhandarya |
| Tupolang | 1985 | 500 | 30 | Tupolang |
| Shurkul | 1978 | 170 | 17 | Zerafshan |
| Kuya- Mazar | 1957 | 320 | 80 | Andizhan Great canal |

| Name | In exploitation since | Total capacity Mm$^3$ | Dead capacity Mm$^3$ | Source |
|---|---|---|---|---|
| Akdarya | 1989 | 130 | 20 | Akdarya |
| Kattakurgan | 1941 | 840 | 24 | Zerafshan |
| Karaultube | 1984 | 53 | 3 | Zerafshan |
| Kamashi | 1957 | 29.5 | 5.7 | Yakkabagdarya |
| Kattasai | 1961 | 55 | 15 | Kattasai |
| Pachkamar | 1967 | 260 | 17 | Guzadarya |
| Dehkanabad | 1983 | 27.2 | 3 | Kychyk-Uradarya |
| Chimkurgan | 1959 | 425 | 0 | Kashkadarya |
| Gissarak | 1982 | 170 | 15 | Aksu |
| Uchkyzyl | 1959 | 160 | 80 | Zang canal |
| | **Total** | **14 464.7** | **3 244.7** | |
| **Turkmenistan** | | | | |
| Zeid | 1963 | 2 200 | 200 | Garagum canal |
| Hauzhan | 1962 | 875 | 25 | Garagum canal |
| Western | 1962 | 48 | 10 | Garagum canal |
| Kopetdag | 1985 | 220 | 25 | Garagum canal |
| Ioloten | 1910 | 73.2 | 1.5 | Murgap |
| Kolhoz-Bent | 1941 | 54.6 | 4.6 | Murgap |
| Kaushutbente | 1895 | 38.2 | 4.5 | Murgap (lower and middle) |
| Sary-Yazin | 1950 | 263 | 15 | Murgap |
| Tedzhen II | 1960 | 183.5 | 3.5 | Tedzhen |
| Tedzhen I | 1950 | 150 | 7.4 | Tedzhen |
| Hor-Hor | 1959 | 21.5 | 0.9 | Tedzhen |
| Mamed-kul | 1964 | 20.5 | 2.5 | Atrek |
| Tashkeprin | 1939 | 166 | 18.3 | Murgap, Kushka |
| | **Total** | **4 313.5** | **3 182** | |
| **Tajikistan** | | | | |
| Muminobod | 1959 | 30.1 | 0.9 | Obisurh |
| Selbur | 1961 | 26 | 0.6 | Kyzylsu |
| Sangtuda | 1980 | 270 | 0 | Vakhsh |
| Golovnaya GES | 1963 | 21.6 | 11 | Vakhsh |
| Baipaza | 1978 | 97 | 13.5 | Vakhsh |
| Nurek | 1970 | 10 500 | 5 964 | Vakhsh |
| | **Total** | **10 944.7** | **5 990** | |
| **Syr-Darya River basin** | | | | |
| **Tajikistan** | | | | |
| Kairakkum | 1956 | 3 413.5 | 894 | Syr-Darya |
| Kattasai | 1961 | 55 | 21.4 | Kattasai |
| | **Total** | **3 468.5** | **915.4** | |
| **Kyrgyzstan** | | | | |
| Toktogul | 1974 | 19 500 | 5 500 | Naryn |
| Uchkurgan | 1901 | 52.5 | 31.6 | Naryn |
| Kurpsai | 1983 | 370 | 20 | Naryn |
| Kurgantepa | 1978 | 33.3 | 5.5 | Shahimardan |
| Naiman | 1971 | 39.5 | 1.5 | Abshirsai, Kyrgyzatan |
| Papan | 1981 | 260 | 10 | Akbura |
| | **Total** | **20 255.3** | **5 568.6** | |

| Name | In exploitation since | Total capacity Mm$^3$ | Dead capacity Mm$^3$ | Source |
|---|---|---|---|---|
| **Uzbekistan** | | | | |
| Dzhizak | 1968 | 100 | 4 | Sanzar |
| Zaamin | 1979 | 51 | 21 | Zaaminsu |
| Charvak | 1966 | 2 000 | 420 | Chirchik |
| Tuyabuguz | 1959 | 250 | 26 | Ahangaran |
| Ahangaran | 1971 | 260 | 30 | Ahangaran |
| Farhad | 1947 | 350 | 330 | Syr-Darya |
| Kassansai | 1942 | 165 | 10 | Kassansai |
| Karkidon | 1963 | 218.4 | 4.4 | Kuvasai and South Fergana canal |
| Andijan | 1978 | 1 900 | 150 | Karadarya |
| | **Total** | **5 294.4** | **995.4** | |
| **Kazakhstan** | | | | |
| Bugun' | 1965 | 350 | 10 | Bugun' |
| Chardara | 1966 | 5 700 | 1 000 | Syr-Darya |
| | **Total** | **6 050** | **1 010** | |
| **Total for Aral Sea basin** | | **64 791.1** | **18 042.3** | |
| **Incl. Amu-Darya** | | **29 722.9** | **9 552.9** | |
| **Syr-Darya** | | **35 068.2** | **8 489.4** | |

**Table 3**

Water use in Uzbekistan (million m³) (SIC ICWC data base, Statistics Yearbooks 10–20)
(*at rayon = district boundaries)

| Year | Source | Drinking water supply | Rural water supply | Industry | Fishery | Irrigation* | Other | Total | Basin |
|---|---|---|---|---|---|---|---|---|---|
| 1960 | | 200 | 250 | 500 | 500 | 13 930 | 0 | 15 380 | Amu |
| | | 280 | 250 | 700 | 200 | 13 970 | 0 | 15 400 | Syr |
| | | 480 | 500 | 1 200 | 700 | 27 900 | 0 | 30 780 | Total |
| 1965 | | 220 | 350 | 600 | 500 | 19 800 | 0 | 21 470 | Amu |
| | | 300 | 350 | 900 | 300 | 17 980 | 0 | 19 830 | Syr |
| | | 520 | 700 | 1 700 | 800 | 37 780 | 0 | 41 500 | Total |
| 1970 | | 250 | 480 | 1 100 | 130 | 23 550 | 0 | 25 510 | Amu |
| | | 450 | 500 | 1 400 | 300 | 19 900 | 0 | 22 550 | Syr |
| | | 700 | 980 | 2 500 | 430 | 43 450 | 0 | 48 060 | Total |
| 1975 | | 300 | 900 | 2 260 | 300 | 25 600 | 100 | 29 460 | Amu |
| | | 500 | 1 050 | 2 600 | 700 | 19 850 | 250 | 24 950 | Syr |
| | | 800 | 1 950 | 4 860 | 1000 | 45 450 | 350 | 54 410 | Total |
| 1980 | | 400 | 1 090 | 2 150 | 200 | 29 060 | 300 | 33 200 | Amu |
| | | 660 | 1 300 | 2 500 | 510 | 26 450 | 290 | 31 710 | Syr |
| | | 1 060 | 2 390 | 4 650 | 710 | 55 510 | 590 | 64 910 | Total |
| 1985 | | 440 | 440 | 240 | 130 | 32 708 | 0 | 33 958 | Amu |
| | | 1 100 | 450 | 1 000 | 290 | 21 596 | 0 | 24 436 | Syr |
| | | 1540 | 890 | 1240 | 420 | 54 304 | 0 | 58 394 | Total |
| 1990 | | 704 | 349 | 196 | 353 | 35 048 | 0 | 36 650 | Amu |
| | | 1 650 | 374 | 1102 | 727 | 23 108 | 0 | 26 961 | Syr |
| | | 2 354 | 723 | 1298 | 1080 | 58 156 | 0 | 63 611 | Total |

| Year | Source | Drinking water supply | Rural water supply | Industry | Fishery | Irrigation* | Other | Total | Basin |
|---|---|---|---|---|---|---|---|---|---|
| 1995 | Total: | 630 | 520 | 250 | 280 | 30 030 | 0 | 31 710 | Amu |
| | Surface water | 0 | 0 | 150 | 280 | 27 230 | 0 | 27 660 | |
| | Ground water | 630 | 520 | 100 | 0 | 1 000 | 0 | 2 250 | |
| | Drainage re-use | 0 | 0 | 0 | 0 | 1 800 | 0 | 1 800 | |
| 1995 | Total: | 1 400 | 570 | 950 | 600 | 18 990 | 0 | 22 510 | Syr |
| | Surface water | 0 | 0 | 300 | 600 | 15 490 | 0 | 16 390 | |
| | Ground water | 1 400 | 570 | 650 | 0 | 1 500 | 0 | 4 120 | |
| | Drainage re-use | 0 | 0 | 0 | 0 | 2 000 | 0 | 2 000 | |
| 1995 | Total: | 2 030 | 1 090 | 1 200 | 880 | 49 020 | 0 | 54 220 | Total |
| | Surface water | 0 | 0 | 450 | 880 | 42 720 | 0 | 44 050 | |
| | Ground water | 2 030 | 1 090 | 750 | 0 | 2 500 | 0 | 6 370 | |
| | Drainage re-use | 0 | 0 | 0 | 0 | 3 800 | 0 | 3 800 | |
| 1999 | Total: | 1 428 | 993 | 527 | 435 | 34 830 | 0 | 38 213 | Amu |
| | Surface water | 0 | 0 | 277 | 435 | 32 255 | 0 | 32 967 | |
| | Ground water | 1 428 | 993 | 250 | 0 | 557 | 0 | 3228 | |
| | Drainage re-use | 0 | 0 | 0 | 0 | 2 018 | 0 | 2 018 | |
| 1999 | Total: | 1 205 | 396 | 817 | 372 | 21 830 | 0 | 24 620 | Syr |
| | Surface water | 0 | 0 | 287 | 372 | 18 096 | 0 | 18755 | |
| | Ground water | 1 205 | 396 | 530 | 0 | 1 600 | 0 | 3 731 | |
| | Drainage re-use | 0 | 0 | 0 | 0 | 2134 | 0 | 2134 | |
| 1999 | Total: | 2 633 | 1 389 | 1 344 | 807 | 56 660 | 0 | 62 833 | Total |
| | Surface water | 0 | 0 | 564 | 807 | 50351 | 0 | 51 722 | |
| | Ground water | 2 633 | 1389 | 780 | 0 | 2 157 | 0 | 6 959 | |
| | Drainage re-use | 0 | 0 | 0 | 0 | 4 152 | 0 | 4 152 | |

**Table 4**
**Main collectors in the Amu-Darya and Syr-Darya river basins, discharge in 1999 (SIC ICWC database)**

| Main collectors | Average flow rate $m^3$/sec (1999) | Salinity g/L | Annual discharge million $m^3$ | Discharged into |
|---|---|---|---|---|
| KS-1 | 12.6 | 4 | 400 | Aral Sea area |
| KS-3 | 5.3 | 4.1 | 168 | Aral Sea area |
| KS-4 | 4.3 | 2.65 | 138 | Aral Sea area |
| KKS | 18.4 | 5.4 | 581 | Sudoche lake |
| Beruny | 10.1 | 3.9 | 321 | Amu-Darya river |
| Ayazkala | 12.4 | 4 | 392 | Ayazkala lake |
| Ustyurt | 6.6 | 4 | 209 | Sudoche lake |
| Ozerny | 73.1 | 4.1 | 2 308 | Sichankul lake |
| Divankul | 31.1 | 2.7 | 981 | Sichankul lake |
| Parsankul | 28.1 | 3 | 884 | Amu-Darya river |
| Dengizkul | 14.3 | 4.9 | 451 | Dengizkul lake |
| Central Bukhara Canal | 15.5 | 3.05 | 490 | Solyonoe lake |
| Zalodno-Romiton | 2.5 | 2.4 | 79 | Solyonoe lake |
| Agitma | 5.5 | 2.1 | 174 | Agitmin lake |
| Severny | 22.6 | 2.9 | 713 | Korakir lake |
| Yuzhny | 74.2 | 7.1 | 2 339 | Sultandag lake |
| Central Golodnaya Steppe Canal | 43.1 | 4.2 | 1 360 | Arnasay depression |
| Shuruzak | 12.7 | 2.8 | 400 | Syr-Darya river |
| DGK | 11.2 | 3.5 | 354 | Kly collector |
| Ok-Bulok | 3.9 | 3.9 | 122 | Tuz kony lake |
| Pogranichny | 1.8 | 4.5 | 58 | Tuz kony lake |
| Kly | 3.23 | 4.7 | 102 | Arnasay depression |
| Achikkul | 49.5 | 2.65 | 1 560 | Syr-Darya river |
| Korakalpak | 9.6 | 1.57 | 302 | Syr-Darya river |
| Sari-Suv | 65.8 | 1.5 | 2 074 | Syr-Darya river |
| Urtukly | 20.6 | 1.3 | 650 | Syr-Darya river |
| Chilisay | 11.2 | 1.27 | 350 | Syr-Darya river |

There has been practically no further irrigation development in Uzbekistan since independence. Management of water resources has focused on operation and maintenance of the existing systems. For instance, in 1994 the capital investment in irrigation development was 37 percent of that of 1990, and only 5 percent in 1999 (Scientific Information Centre of the Interstate Coordination Water Commission – SIC of ICWC). At the same time funds for the operation and maintenance of irrigation systems in 1999 decreased by 32 percent compared to their level in 1990. In 1999, the operating costs of water resources organizations of Uzbekistan were US$436.1 million, or US$103.8 per ha.

In the region, i.e. Central Asia, Uzbekistan is the largest water user with the least potential to generate water resources. There is therefore a great need to overcome water deficit and to solve the problems of transboundary water resources management. Old principles of water management which were applied to the whole region prior to independence, giving priority to irrigated agriculture, are not in accordance with the priority of water use as a basic source of power generation in the independent states located in the upper watershed area (Kyrgyzstan and Tajikistan). This leads to a conflict of interests between upstream and downstream states. To develop cooperation in the field of joint management, protection and use of water resources, the regional ICWC with its executive bodies, the Basin Water Organisations (BWO) Amu-Darya and Syr-Darya, and the Scientific Information Centre of ICWC were established in 1992 by the initiative of the ministers of five Central Asian states. The present structure of water resources management in the region consists of two levels, i.e. interstate and national.

The present national institutional setting responsible for irrigation development and management in Uzbekistan consists of a number of departments financed by the government.

The main constraints to irrigation development in Uzbekistan are:

- inadequate system of centralized command management of land and water resources;
- insufficient financing;
- rising water deficit;
- conflict of water development interests between upstream and downstream countries in the Amu-Darya and Syr-Darya river basins;
- physical and functional deterioration of water assets;
- migration of qualified personnel due to the decreasing living standard;
- environmental problems caused by deterioration of water quality, Aral Sea desiccation and desertification of Priaralye (areas around the Aral Sea).

## 1.2 Status of the use of irrigation systems for fish production

Until 1960 fisheries of Uzbekistan were concentrated predominantly in the Aral Sea area. In 1958 fish catches reached a maximum of 50 000 tonnes (Fig. 1). As a result of the Aral Sea desiccation and increased salinity to 14 g/litre (the salinity in 1983), there has been no fishing in the Aral Sea since 1983. Fisheries have moved to delta lakes and Lake Sarykamysh formed from drainage waters. But Sarykamysh also eventually lost its fishery value due to an increase in salinity which by now reached in some areas 20 g/litre. Today's fish yields in lakes and reservoirs in the Aral Sea area range from 1.2 to 209 kg/ha (Anon., 1990, 1998, 2001a) (Table 5).

**Fig. 1**

Fisheries in the Aral Sea zone (Zholdasova *et al.*, 1996)

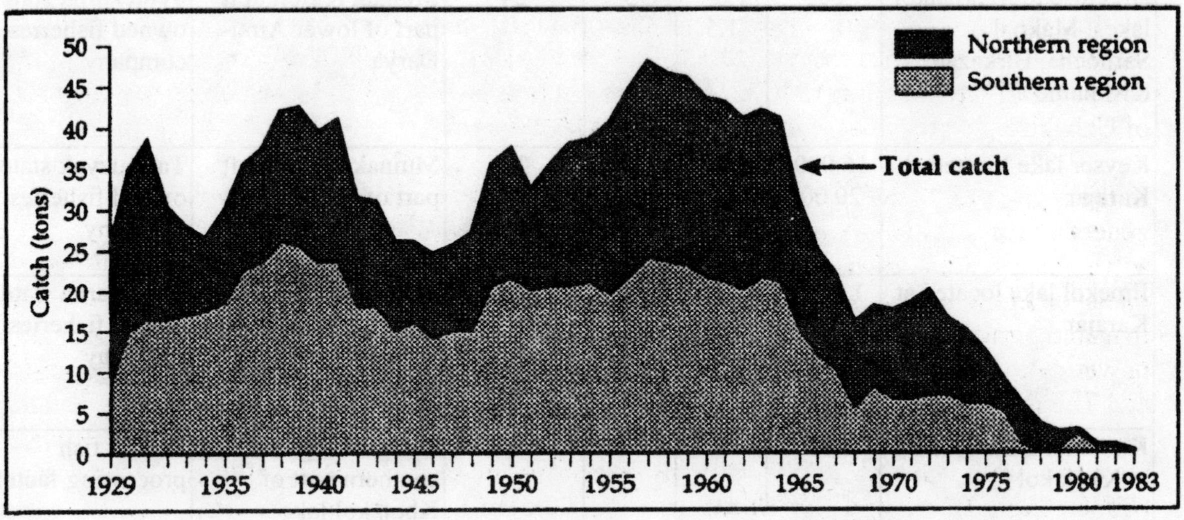


**Table 5**

Fish production in lakes and reservoirs owned by the State Joint Stock Company Karakalpak Fisheries (Anon., 2001a)

| Lake or reservoir | Area (ha) | Depth (m) | Width (m) | Production kg/ha | Reservoir location | Name of fishing enterprise |
|---|---|---|---|---|---|---|
| Domalak, Janly – close to Domalak lakes | 2 000 | 1.2 – 1.5 | 3.5 – 4.0 | 80 | Muinak region right bank of Amu-Darya | Muinak fish processing factory |
| Karateren lake – located at Damalike | 1 000 | 1.2 – 1.5 | 3.5 – 4.0 | 40 | Muinak region right bank of Amu-Darya | Kazakh-Darya fish processing factory |
| Shege | 3 000 | 1.3 – 1.8 | 3.5 | 66 | Muinak region, northwestern part of Mezhdurechye | Amu-Darya state-owned fisheries company |
| Kok-suu | 1 500 – 2 500 | 1.3 – 1.8 | 25 | 40 | | |
| Sudachye including Ak-ushpa, Taily and Urge lakes | 3 300 | 0.7 – 0.8 | 1.7 | 59 | Northeastern part of the left bank of Amu-Darya near GLK collector in the west of Sudachye lake | Uchsay fish processing factory |
| Large Sudachye lake | | | | | Near GLK collector in the east part of lake Sudachye | Konyrat fish processing factory |

| Lake or reservoir | Area (ha) | Depth (m) | Width (m) | Production kg/ha | Reservoir location | Name of fishing enterprise |
|---|---|---|---|---|---|---|
| All close to Makpalkol lakes: Makpal, Sarhocha, Birkazan Kisilkeme | 600 | 1.0 – 1.5 | 3 | 24 | Muinak region, left part of lower Amu-Darya | Tentekarna state-owned fisheries company |
| Keyser lake located at Karajar | 16 000–20 000 | 2.0 – 2.2 1.0 – 1.2 | 3 3 | 39 88 | Muinak region, left part of Amu-Darya | Taly-uzyak state-owned fisheries company |
| Ilmekol lake located at Karajar | 1 000 – 1 500 | | | | Muinak region, central part of Amu-Darya left bank near Karakhar | Tentekarna state-owned fisheries company |
| Khojahol Lake located at Khojakol | 1 000 | 2 | 3.5 | 55 | Kungrad region, southern part of Khojakol lake | Konrat fish processing factory |
| Koptin-kol Lake located at Khojakol | 9 500 | 1.2 – 2.0 | 6 | 209 | Kungrad region, central part of Amu-Darya left bank and near Khojakol lake | Konrat fish processing factory |
| Jaunger-kol Lake located at Khojakol | 532 | 1.7 – 1.8 | 2.5 – 3.0 | 8 | Kungrad region, central part of Amu-Darya left bank and near Khojakol lake | Konrat fish processing factory |
| Muinak gulf | 9 750 | 1.65 | 3 | 78 | Muinak region, southeastern part of the Aral Sea | Nurly-hol fisheries cooperative |
| Rybachy gulf (Sarybas) | 4000 | 1.0 – 1.5 | 3.4 | 34 | Munaik region in 1 km to the North of Muinak in the southern part of dried bed of the Aral Sea | Tentek Arna fish processing factory |
| Sarykamysh | 300 000 | 5.7 | 47 | 33 | South-west of the Aral Sea, 200 km from Turkmenistan border with Karakalpakstan | |
| Shygys Carateren | 4 000 | 3 – 5 | 30 | 52 | Tahtakul region, foot of Beltau Height | Tahtakul fishery |

| Lake or reservoir | Area (ha) | Depth (m) | Width (m) | Production kg/ha | Reservoir location | Name of fishing enterprise |
|---|---|---|---|---|---|---|
| Botakol and nearby lakes | 2 000 | 1.5 – 2.0 | 4 | 18 | Tahtakulsky region between Karateren lake and Kok-Darya River (KS – 4) | |
| Atakol and nearby lakes | 2 000 | 1.5 – 2.0 | 3.5 – 4.5 | 10 | Tahtakulsky region | |
| Tashpenkol | 1 000 | 1.5 – 3 | 10 | 1.2 | Wimbaisk region, Amudarya right bank, Kuskanatau Height | Wimbaisk fishery |
| Dauytkol reservoir | 5 000 | 1.5 – 2.0 | 7 | 80 | Amu-Darya right bank, 47 km to the north of Nukus | Nukus fish processing factory |
| Karakol | 7 000 | 0.9 – 2.6 | 2.6 | 115 | Northwestern part of Shumanai region | Shumanai fishery |
| Akshakol | 4 000 | 1.5– 20.0 | 7 | 20 | Ellicalin region, Amu-Darya right side, 20–30 km from southeastern part of Sultanuzdag | Ellikalin fishery |
| Ayazkala | 9 000 | 2 – 3 | 8 | 40 | Berunian region, Amu-Darya right bank and southern part of Sultanuzdag | Berunina fishery |
| Zhyltyrbas | 30 000 | | | | Muinak region | Kasakhdarya fish processing factory |
| Kobeyshungil, Sarykol, Magnit zhargan | 50 | 1.0– 1.5 | 2 | 0.3 | Karauzyak region | Karauzyak fishery |

When Uzbekistan was part of the Soviet Union, the Ministry of Fisheries of the former USSR, in cooperation with the Ministry of Water Management and Uzbekistan Government, in the 1960s–1970s developed a large-scale comprehensive programme of fish production for all types of inland waterbodies. Special attention was paid to education, research, planning, water and fish quality monitoring and other issues. Due to the presence of irrigation systems throughout the plains of Uzbekistan, fisheries planners, developers and managers had to make the best use of different types of irrigation waterbodies (reservoirs, irrigation and drainage canals, lakes storing drainage water).

State-owned fishing companies were established at all large reservoirs and lakes for return water storage. Hatcheries for producing stocking material were constructed in all parts of Uzbekistan. By the 1980s up to 7 000 tonnes of fish per year were harvested from reservoirs and lakes. All fish farms were state owned, financed by the government, and functioned within the structure of the Ministry of Fisheries. They regularly reported on their fish production.

After the dissolution of the USSR the situation changed significantly. The government of Uzbekistan privatized all state-owned fish farms and capture fisheries enterprises and stopped their government financial support. As a result fish production dropped to one third and large-scale fishing in reservoirs such as Charvak, Chimkurgan and several others virtually stopped or was significantly reduced, as for example in Tudakul reservoir, where the reported fish catches dropped from 700 tonnes in the early 1990s to 250 tonnes in the late 1990s. Fish production in ponds has decreased on the average from 3 000 kg to 850 kg per ha. Education and training of specialists also stopped, and the research network came to an almost complete standstill (Kamilov, this volume).

Today, Uzbekistan has no national programme or specific fishery development projects supported by the government or international assistance. Private initiative focuses only on exploitation of rich fish stocks in the Aidaro-Arnasay system using small fishing teams. The fishery potential of waterbodies of the irrigation system of Uzbekistan is largely unexploited.

Between 1996 and 2001 two project proposals were formulated for the development of fisheries on fish farms of the enterprise "Uzbalik". One, a project for a model aquaculture farm using semi-intensive technology, was prepared in cooperation with the Ministry of Foreign Affairs of Israel; the other, for a fish farm for sturgeon production, was prepared in cooperation with a German company. Both projects have not yet been implemented, mainly for the lack of funds on the Uzbek side and because of insufficient experience of Uzbek fishery specialists in implementing such projects.

Prior to independence some experience was obtained in using irrigation systems for fish production in the Golodnaya and Karshy steppes. In the Southern Golodnaya Steppe Canal an experimental fish hatchery for grass carp (*Ctenopharyngodon idella*) and silver carp (*Hypophthalmichthys molitrix*) was constructed and operated during 1975–1985. This hatchery produced fingerlings for stocking irrigation canals overgrown with aquatic macrophytes. Fish stocked into the canals cleared them almost completely of aquatic plants and the fish themselves were afterwards harvested for food. While this approach is considered by fishery specialists and engineers as an efficient aquatic weed controlling mechanism, the hatchery ceased functioning and canals are now again overgrowing with aquatic plants. In Karshy steppe there was fisheries in Talimarjan reservoir, and fish ponds were constructed to use drainage waters of Sichankul and Sultan Uezdak. These ponds became successful in fish production. Even this undertaking is now largely neglected.

## 2. POTENTIAL FOR THE USE OF IRRIGATION SYSTEMS FOR FISH PRODUCTION

Irrigation systems are present throughout Uzbekistan. Uzbekistan has a good transportation and industrial infrastructure, large rural population and diversified agriculture. All this creates favourable social and economic conditions for development of fish production in all waterbodies connected with irrigation.

Uzbekistan is located in a climatic zone favourable for development of aquaculture in natural waterbodies and in reservoirs (Fig. 2). There is a long season of warm and sunny weather which keeps water warm. The existing water resources provide for a large increase in both capture fisheries and aquaculture. The presence of cold waters at a higher altitude enables production of cold water fish, including trout.

The lowland reservoirs of the Zaravshan, Kashka-Darya, Surkhan-Darya basins of the Amu-Darya middle reaches are suitable for the development of capture fisheries based on warmwater Chinese carps, i.e. silver, bighead, and grass carp, and common carp. In the mountain and foothill storage reservoirs such as Charvak, Akhangaran, Andizhan and several others one could produce cold water fish such as rainbow trout, Issyk-Kul (Sevan) salmon and whitefish (*Coregonus* spp.). It would be profitable to use the existing ponds, now in private ownership, for creation of fish farms. With a model of a profitable small fish farm, farmers of Uzbekistan could then combine pond fish culture with the traditional farm crop production. This would appear to be an efficient way of boosting the fish production in the country.

Fish hatcheries in Uzbekistan are still functioning, including the breeding and production of fish fry and fingerlings of cyprinids. They have sufficient capacity to provide potential farmers with the required quantities of fry, fingerlings and yearlings for increasing fish production to up to 100 000 tonnes per year. Prior to independence the Uzbekistan fish farms supplied all Central Asia, southern Siberia and some other regions with fry and fingerlings (Kamilov, this volume).

Today there is an obvious shortage of fish in Uzbekistan, with a fish supply of less than 1 kg/person/year. Before 1990 the country processed and consumed annually 70–100 000 tonnes of fish, of which 30 000 tonnes were from the local production, and 40–70 000 tonnes imported. Today's consumption is only 9–10 000 tonnes (Kamilov, this volume).

The following conditions favour further development of fisheries in Uzbekistan:

- social – optimum density of population and need to create new workplaces;
- economic – proximity to markets, especially to those with traditionally high fish demand; proximity to large industrial centres, which would facilitate introduction of new technologies for production and marketing of high value fish species;
- climatic – high level of solar radiation, diversity of climatic zones favouring the production of both warm water and cold water fish species;
- technological – availability of both irrigation and drainage canals-collectors for fish production (Table 6).

**Fig. 2**
**Climatic zones in the Aral Sea basin**

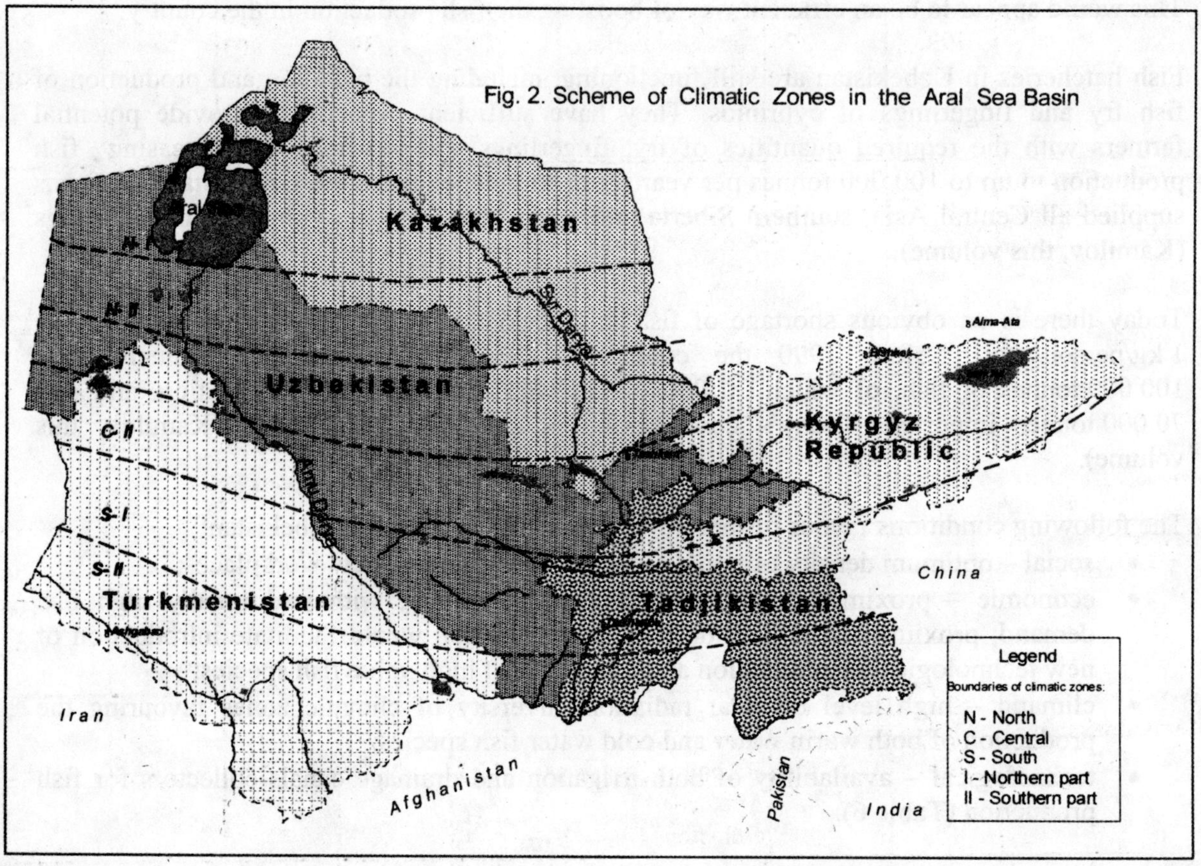

**Table 6**

Return waters in the Aral Sea basin (SIC ICWC data base)

Return waters in the Amu-Darya River basin (Uzbekistan) (million $m^3$ = $Mm^3$)

| Catchment areas, oblasts | Years of observ-ations | Catch-ment area, ('000 ha) | Main tributaries | Average salinity g/litre | Return waters ($Mm^3$) | Diverted | | |
|---|---|---|---|---|---|---|---|---|
| | | | | | | to a river ($Mm^3$) | to depres-sions ($Mm^3$) | for irrig-ation ($Mm^3$) |
| **Upper reaches** | | | | | | | | |
| Surkhandarya | 1990 | 311 | Karasu river | 2.35 | 1 138 | 510 | – | 628 |
| | 1995 | 324 | Antorsky collect. | 1.55 | 1 398 | 803 | – | 595 |
| | 1999 | 320 | VST, K-1, K-2, …K-5 | 2.2 | 1 100 | 600 | – | 500 |
| **Middle reaches** | | | | | | | | |
| a) Kashkadarya | 1990 | 485 | Yuzhny collector | 6.76 | 1 722 | 654 | 1 068 | – |
| | 1995 | 495 | Kashkadarya river | 5.74 | 2 043 | 650 | 1 393 | – |
| | 1999 | 490 | | 7.1 | 2 340 | 800 | 1 540 | – |
| b) Bukhara | 1990 | 343 | Parsankul, Central | 3.69 | 2 273 | 650 | 1 623 | – |
| | 1995 | 390 | Bukhara Canal | 3.11 | 1 938 | 510 | 1 428 | – |
| | 1999 | 385 | Severny collector Dengizkul | 4.2 | 2 790 | 810 | 1 980 | – |
| **Lower reaches** | | | Ozerny collector | | | | | |
| a) Khorezm | 1990 | 258 | Divankul | 2.98 | 2 740 | – | 2 561 | 179 |
| | 1995 | 267 | collector | 2.02 | 4 009 | – | 3 819 | 190 |
| | 1999 | 256 | | 3.7 | 3 290 | – | 3 110 | 180 |
| b) Karakalpakstan | 1900 | 495 | KS-1, KS-3 | 4.2 | 2 332 | 388 | 1 944 | – |
| | 1905 | 500 | KS-4, KKS, | 3.41 | 2 897 | 370 | 2 527 | – |
| | 1999 | 500 | Beruny collector Ayazkala collector | 4.2 | 2 200 | 320 | 1 880 | – |
| **Total in the basin ($Mm^3$)** | 1990 | 1892 | | 3.98 | 10 205 | 2 202 | 7 196 | 807 |
| | 1995 | 1976 | | 3.08 | 12 285 | 2 333 | 9 167 | 785 |
| | 1999 | 1951 | | 4.45 | 11 720 | 2 530 | 8 510 | 680 |

Return waters in the Syr-Darya River basin (Uzbekistan)

| Catchment areas, oblasts | Years of observ-ations | Catchment area ('000 ha) | Main tributaries | Average salinity g/litre | Return waters ($Mm^3$) | Diverted | | |
|---|---|---|---|---|---|---|---|---|
| | | | | | | to a river ($Mm^3$) | to depres-sions ($Mm^3$) | for irrig-ation ($Mm^3$) |
| Andizhan | 1990 | 277 | Zambarkul | 1.21 | 2523 | 2269 | – | 254 |
| | 1995 | 265 | Karagukon | 1.46 | 2710 | 2440 | – | 270 |
| | 1999 | 280 | | 1.65 | 1264 | 1200 | – | 64 |
| Namangan | 1990 | 268 | Achikkul | 2.21 | | 1010 | – | 1050 |
| | 1995 | 276 | Karakalpok | 1.08 | 2037 | 1030 | – | 1007 |
| | 1999 | 280 | | 2.75 | 2250 | 1090 | – | 1160 |
| Ferghana | 1990 | 3 | Achikkul | 2.21 | 3050 | 2180 | – | 870 |
| | 1995 | 4 | Saridzhuga | 2.28 | 3300 | 2400 | – | 900 |
| | 1999 | 9 | Severny | 2.8 | 2970 | 2020 | – | 950 |
| Tashkent | 1990 | 379 | Urtukly | 1.45 | 2569 | 2329 | – | 240 |
| | 1995 | 397 | Karasuv | 1.29 | 2135 | 1931 | – | 204 |
| | 1999 | 390 | Chilisoy | 2.1 | 2480 | 2200 | – | 280 |
| Syrdarya | 1990 | 298 | Central | 4.19 | 1712 | 743 | 876 | 93 |
| | 1995 | 298 | Golodnaya | 3.49 | 1571 | 831 | 670 | 70 |
| | 1999 | 280 | Steppe Canal Shuruzyak | 3.6 | 1940 | 390 | 1450 | 100 |
| Djizak | 1990 | 2 | Djizak main canal, Ok-Bulak KLI | 5.6 | 778 | – | 736 | 42 |
| | 1995 | 8 | | 4.41 | 1318 | | 1218 | 100 |
| | 1999 | 5 | | 44 | 1165 | | 1100 | 65 |
| **Total in the basin ($Mm^3$)** | 1990 | 1856 | | 2 | 12692 | 8531 | 1612 | 2549 |
| | 1995 | 1885 | | 3 | 13071 | 8632 | 1888 | 2551 |
| | 1999 | 1890 | | 3 | 12069 | 6900 | 2550 | 2619 |

Karakalpakstan Republic and Navoy, Samarkand and Djizak oblasts (provinces), continue to capture a relatively large amount of fish from reservoirs of the lower Amu-Darya and the Aidar-Arnasay system of lakes. The densely populated Fergana valley, has a severe fish deficit.

Taking into account the above conditions, the following areas of Uzbekistan appear to have a good potential for the use of irrigation systems for fish production:

| Favourable conditions | Areas with good fish production potential in irrigation systems |
|---|---|
| Social | Namangan, Andizhan, Fergana, Tashkent and Samarkand oblasts |
| Economic | Khorezm, Bukhara, Kashkadarya and Tashkent oblasts;<br>to some degree: Karakalpakstan Republic, Navoy, Samarkand and Djizak oblasts |
| Institutional | Tashkent, Samarkand, Bukhara and Khorezm oblasts |
| Climatic | Fergana valley, Tashkent, Samarkand, Syr-Darya, Djizak and Khorezm oblasts |
| Technological | all oblasts of Uzbekistan |

# 3. CONSTRAINTS IN USING IRRIGATION SYSTEMS FOR FISH PRODUCTION

Difficulties of maintaining and developing irrigation in Uzbekistan negatively affect fish production. Among the factors limiting the use of irrigation systems for fish production are the following:

*Institutional constraints*. Lack of governmental and non-governmental institutional structures to promote the use of irrigation systems for fish production. Absence of legislation ensuring the rights of private fish farmers to a guaranteed water supply within special limits and to trade in fish.

*Economic constraints*. Lack of government financing and private investments in the industry. Absence of specialized credit lines.

*Technical constraints*. Lack of protecting devises on diversion structures which would prevent young fish from being discharged with irrigation water onto irrigated fields; lack of corridors between waterbodies including floodplains, river reaches and canals, to make possible the migration of fish and fish fry from and to places of spawning, reproduction and other types of existence; unsuitable or absence of fish passes; priorities for water use, i.e. irrigation demand and hydropower production, which often do not allow maintaining optimal water supply for fish spawning and in nursery grounds.

*Ecological constraints*: water pollution in irrigation systems, including increased salinities and toxicity.

*Social and cultural constraints*. Low level of public awareness that the irrigation network can be used for fish production. Shortage of fisheries experts and of fisheries training programmes.

Some constraints cannot be resolved, therefore fish production should be developed in view of the following constraints: (i) water is regulated only for the needs of irrigated agriculture and/or hydropower production, without inclusion of fishery management requirements for sustainable fish production from irrigation waterbodies; (ii) irrigation systems in Uzbekistan are owned by the government, managed by the government and financed from the state budget. The very low budgetary allocation is slowing down any attempt to use irrigation systems for a sound development of fish production.

The following constraints need urgent attention:

- low awareness on the part of central and local authorities of the importance of fish in nutrition of people, and of the potential and economic profitability of fish production. This could be resolved by providing cheap credits and supporting private initiatives;
- absence of fish protection devices on water intakes from rivers, canals and reservoirs; the technology is available, simple, and not expensive;
- old, no longer appropriate fisheries laws and regulation;
- all fish in waterbodies of Uzbekistan, except for fish produced in farm ponds, is state property, including fish in the canals managed by water users;
- there are too many government authorities of various levels on which the development of fisheries depends: Ministry of Agriculture and Water Management is in charge of waterbody uses; State Committee of the Environment issues fishing permits; fish landings must be reported to the local authorities;
- there is no appropriate body or laws protecting the interests of fisheries. For example, in case of water pollution the damage is measured and levied to the state budget, but fishermen are not compensated for fish losses. Formerly assistance was provided by the Fishery Inspection, which does not exist any more and no alternative organ has been established to carry out its duties;
- rise in illegal fishing due to an insufficient government attention being given to fisheries; illegal fishermen are now one of the major fish suppliers to the markets;
- in recent years water shortage has limited fish production in the existing fish farms with large ponds; this shows the need for focusing in the future on the use of small ponds and water-saving technologies, which would also be a more suitable approach under the new market economy conditions.

The following constraints may be difficult to address, at least initially. These constraints seem to apply also to other countries of Central Asia:

- lack of economical and technological models for the development of private fisheries in irrigation systems;
- lack of financial support for scientific and applied research in this direction;
- lack of experience in obtaining credits for establishment of fish farms and in attracting foreign participation for private ventures;
- lack of international assistance for the development of fish production.

# 4. RECOMMENDATIONS FOR BETTER USE OF IRRIGATION SYSTEMS FOR FISH PRODUCTION

The experience from the period 1960–1980 shows that in Uzbekistan a significant amount of fish can be produced in irrigation systems (canals, reservoirs, drainage collectors, lakes accumulating drainage water). But at present the irrigation systems are hardly used for this purpose as former mechanisms of fisheries management have disappeared, and new mechanisms have not been created. The rehabilitation of fisheries and further progress in this direction will require a number of measures to be taken:

**Institutional aspects:**

- taking into account the institutional integrity of agriculture and water management, establish within the Ministry of Agriculture and Water Resources of Uzbekistan a department for the use of irrigation systems for fish production;
- create conditions for involving in fish production the personnel operating and maintaining irrigation systems; this would provide them with additional income and also solve the problem of a rapid employee turnover;
- develop a system of public awareness through mass media;
- develop the legislative base, precisely setting out the rights and duties of fish producers and protecting their interests;
- consider accepting the following institutional framework for fish production: Department of Fisheries under the aegis of the Ministry of Agriculture and Water Resources, Agricultural research organization, Aquaculture associations.

**Economic aspects:**

- utilize favourable conditions arising from on-going reforms in agriculture and water management for development of private small-scale fish businesses in the form of aquaculture in small ponds. Institutionally it could be done within the Associations of Farmers and Water-users.

**Technical, training and research aspects:**

- develop a programme of regular water releases to prevent salinization of waterbodies having good quality water, or to prevent increase in salinity of waterbodies which are already saline, such as Sarykamysh; construct fish protection devices on intake structures; consider the need of fish producers for seasonal flow regulation;
- initiate subsequent or simultaneous water use for several purposes, for instance for combining irrigation and drainage with fisheries;
- develop strategies for rational use for fish production of irregular and untimely water releases and flood waters;
- establish an experimental research centre for development of fishery technologies for use in irrigation systems; the centre would also monitor global development and trends, identify the most appropriate technologies and adapt them for the local geographical, social and economic conditions;
- test selected technologies in pilot projects, with the objective of applying them throughout Uzbekistan and in other countries of Central Asia;
- develop and establish training courses in aquaculture which would be run by the ICWC Training Centre, Tashkent, Uzbekistan.

# 5. REGIONAL SITUATION

Fisheries development in Central Asia has a good potential given the favourable climatic conditions prevailing in this geographical area. This is also supported by the favourable socio-economic setup, with abundance of labour source. The rapid population growth means also increasing demand for food, and a further development and expansion of fisheries is one of the ways which to go.

The available data on water use country by country has shown a significant difference in the use of the total flow for the needs of fisheries (Table 3 and Table 7). For instance, in 1999, 807 million m³ of water or 1.3 percent of its total amount were used for fisheries in Uzbekistan, 61 million m³ or 1.1 percent in Kazakhstan, 75 million m³ or 0.7 percent in Tajikistan, 23 million m³ or 0.01 percent in Turkmenistan and almost zero in Kyrgyzstan. One may conclude that Uzbekistan of all countries in the Region has the best potential for using irrigation systems for fish production.

**Table 7**

**Water use in countries of Central Asia (SIC ICWC data base, Statistics Yearbooks 10–20)**

**Water use in Kazakhstan (within the Aral Sea basin) (million m³)**

| Year | Source | Drinking water supply | Rural water supply | Industry | Fishery | Irrigation (at rayon boundaries) | Other | Total |
|---|---|---|---|---|---|---|---|---|
| 1960 | | 65 | 50 | 100 | 40 | 9 495 | 0 | 9 750 |
| 1965 | | 75 | 60 | 120 | 80 | 10 465 | 0 | 10 800 |
| 1970 | | 95 | 90 | 170 | 150 | 12 275 | 70 | 12 850 |
| 1975 | | 105 | 105 | 200 | 300 | 11 400 | 100 | 12 210 |
| 1980 | | 120 | 130 | 220 | 450 | 12 830 | 150 | 14 200 |
| 1985 | | 126 | 143 | 260 | 590 | 9 736 | 160 | 11 015 |
| 1990 | | 214 | 147 | 276 | 111 | 10 136 | 437 | 11 320 |
| 1995 | Total: | 140 | 130 | 180 | 150 | 10 100 | 600 | 11 300 |
| | Surface water | 0 | 30 | 60 | 150 | 10 000 | 285 | 10 525 |
| | Ground water | 140 | 100 | 120 | 0 | 0 | 15 | 375 |
| | Drainage re-use | 0 | 0 | 0 | 0 | 100 | 300 | 400 |
| 1999 | Total: | 54 | 51 | 53 | 61 | 4 701 | 327 | 5 247 |
| | Surface water | 0 | 0 | 0 | 61 | 4 666 | 112 | 4 839 |
| | Ground water | 54 | 51 | 53 | 0 | 0 | 15 | 173 |
| | Drainage re-use | 0 | 0 | 0 | 0 | 35 | 200 | 235 |

**Water use in Kyrgyzstan (within the Aral Sea basin) (million m³)**

| Year | Source | Drinking water supply | Rural water supply | Industry | Fishery | Irrigation (at rayon boundaries) | Other | Total |
|---|---|---|---|---|---|---|---|---|
| 1960 | | 40 | 25 | 28 | 0 | 2 117 | 0 | 2 210 |
| 1965 | | 45 | 35 | 31 | 0 | 2 629 | 0 | 2 740 |
| 1970 | | 55 | 40 | 35 | 0 | 2 850 | 0 | 2 980 |
| 1975 | | 70 | 45 | 44 | 0 | 3 411 | 0 | 3 570 |
| 1980 | | 75 | 50 | 50 | 10 | 3 895 | 0 | 4 080 |
| 1985 | | 79 | 52 | 58 | 11 | 4 070 | 0 | 4 270 |
| 1990 | | 94 | 70 | 68 | 13 | 4 910 | 0 | 5 155 |
| 1995 | Total: | 91 | 85 | 56 | 5 | 4 730 | 0 | 4 966 |
| | Surface water | 48 | 20 | 6 | 5 | 4 470 | 0 | 4 549 |
| | Ground water | 43 | 65 | 50 | 0 | 176 | 0 | 334 |
| | Drainage re-use | 0 | 0 | 0 | 0 | 84 | 0 | 84 |

| Year | Source | Drinking water supply | Rural water supply | Industry | Fishery | Irrigation (at rayon boundaries) | Other | Total |
|---|---|---|---|---|---|---|---|---|
| 1999 | Total: | 90 | 52 | 44 | 0 | 3 100 | 5 | 3 291 |
| | Surface water | 45 | 0 | 0 | 0 | 2 897 | 0 | 2 942 |
| | Ground water | 45 | 52 | 44 | 0 | 150 | 0 | 291 |
| | Drainage re-use | 0 | 0 | 0 | 0 | 53 | 5 | 58 |

Water use in Tajikistan (million $m^3$)

| Year | Source | Drinking water supply | Rural water supply | Industry | Fishery | Irrigation* | Other | Total | Basin |
|---|---|---|---|---|---|---|---|---|---|
| 1960 | | 150 | 250 | 150 | 50 | 6 670 | 250 | 7 520 | Amu |
| | | 100 | 90 | 40 | 30 | 2 020 | 0 | 2 280 | Syr |
| | | 250 | 340 | 190 | 80 | 8 690 | 250 | 9 800 | Total |
| 1965 | | 200 | 400 | 250 | 70 | 7 210 | 400 | 8 530 | Amu |
| | | 130 | 110 | 60 | 50 | 2 320 | 0 | 2 670 | Syr |
| | | 330 | 510 | 310 | 120 | 9 530 | 400 | 11 200 | Total |
| 1970 | | 300 | 580 | 320 | 100 | 8 640 | 500 | 10 440 | Amu |
| | | 150 | 120 | 80 | 80 | 2 530 | 0 | 2 960 | Syr |
| | | 450 | 700 | 400 | 180 | 11 170 | 500 | 13 400 | Total |
| 1975 | | 390 | 640 | 410 | 120 | 7 850 | 600 | 10 300 | Amu |
| | | 145 | 150 | 75 | 100 | 3 330 | 0 | 3 800 | Syr |
| | | 535 | 790 | 485 | 220 | 11 180 | 600 | 14 100 | Total |
| 1980 | | 400 | 650 | 420 | 130 | 8 550 | 600 | 1 0750 | Amu |
| | | 150 | 160 | 70 | 100 | 3 270 | 0 | 3750 | Syr |
| | | 550 | 810 | 490 | 230 | 11 820 | 600 | 14 500 | Total |
| 1985 | | 350 | 620 | 400 | 120 | 7 771 | 552 | 9 813 | Amu |
| | | 142 | 155 | 65 | 96 | 2 142 | 0 | 2 600 | Syr |
| | | 492 | 775 | 465 | 216 | 9 913 | 552 | 12 413 | Total |
| 1990 | | 309 | 563 | 537 | 336 | 7 140 | 374 | 9 259 | Amu |
| | | 176 | 133 | 57 | 123 | 3 099 | 0 | 3 588 | Syr |
| | | 485 | 696 | 594 | 459 | 10 239 | 374 | 12 847 | Total |
| 1995 | Total: | 320 | 402 | 400 | 90 | 8 200 | 40 | 9 452 | |
| | Surface water | 170 | 344 | 349 | 90 | 7 920 | 20 | 8 891 | Amu |
| | Ground water | 150 | 60 | 51 | 0 | 200 | 20 | 481 | |
| | Drainage reuse | 0 | 0 | 0 | 0 | 80 | 0 | 80 | |
| 1995 | Total: | 155 | 183 | 49 | 50 | 2 200 | 0 | 2 637 | |
| | Surface water | 5 | 93 | 0 | 50 | 1 510 | 0 | 1 658 | Syr |
| | Ground water | 150 | 90 | 49 | 0 | 400 | 0 | 689 | |
| | Drainage re-use | 0 | 0 | 0 | 0 | 290 | 0 | 290 | |
| 1995 | Total: | 475 | 585 | 449 | 140 | 10 400 | 40 | 12 089 | |
| | Surface water | 175 | 435 | 349 | 140 | 9 430 | 20 | 10 549 | Total |
| | Ground water | 300 | 150 | 100 | 0 | 600 | 20 | 1 170 | |
| | Drainage reuse | 0 | 0 | 0 | 0 | 370 | 0 | 370 | |
| 1999 | Total: | 490 | 135 | 370 | 50 | 7 050 | 800 | 8 895 | |
| | Surface water | 320 | 100 | 330 | 50 | 6 830 | 785 | 8 415 | Amu |
| | Ground water | 170 | 35 | 40 | 0 | 150 | 15 | 410 | |
| | Drainage re-use | 0 | 0 | 0 | 0 | 70 | 0 | 70 | |

| Year | Source | Drinking water supply | Rural water supply | Industry | Fishery | Irrigation* | Other | Total | Basin |
|---|---|---|---|---|---|---|---|---|---|
| 1999 | Total: | 250 | 55 | 50 | 25 | 2 000 | 150 | 2 530 | Syr |
| | Surface water | 90 | 0 | 0 | 25 | 1 490 | 150 | 1 755 | |
| | Ground water | 160 | 55 | 50 | 0 | 300 | 0 | 565 | |
| | Drainage re-use | 0 | 0 | 0 | 0 | 210 | 0 | 210 | |
| 1999 | Total: | 740 | 190 | 420 | 75 | 9 050 | 950 | 11 425 | Total |
| | Surface water | 410 | 100 | 330 | 75 | 8 320 | 935 | 10 170 | |
| | Ground water | 330 | 90 | 90 | 0 | 450 | 15 | 975 | |
| | Drainage re-use | 0 | 0 | 0 | 0 | 280 | 0 | 280 | |

(*) at rayon boundaries

Water use in Turkmenistan (million m$^3$)

| Year | Source | Drinking water supply | Rural water supply | Industry | Fishery | Irrigation* | Other | Total |
|---|---|---|---|---|---|---|---|---|
| 1960 | | 60 | 25 | 25 | 10 | 7 950 | 0 | 8 070 |
| 1965 | | 80 | 30 | 30 | 15 | 11 345 | 0 | 11 500 |
| 1970 | | 90 | 35 | 35 | 18 | 17 092 | 0 | 17 270 |
| 1975 | | 110 | 40 | 50 | 10 | 22 630 | 0 | 22 840 |
| 1980 | | 130 | 60 | 60 | 15 | 22 735 | 0 | 23 000 |
| 1985 | | 146 | 80 | 135 | 28 | 24 571 | 0 | 24 960 |
| 1990 | | 187 | 42 | 111 | 35 | 22 963 | 0 | 23 338 |
| 1995 | Total: | 330 | 70 | 325 | 35 | 22 470 | 0 | 23 230 |
| | Surface water | 145 | 40 | 289 | 35 | 22 274 | 0 | 22 783 |
| | Ground water | 185 | 30 | 36 | 0 | 151 | 0 | 402 |
| | Drainage re-use | 0 | 0 | 0 | 0 | 45 | 0 | 45 |
| 1999 | Total: | 363 | 188 | 710 | 23 | 16 788 | 3 | 18 075 |
| | Surface water | 173 | 148 | 670 | 23 | 16 601 | 3 | 17 618 |
| | Ground water | 190 | 40 | 40 | 0 | 150 | 0 | 420 |
| | Drainage re-use | 0 | 0 | 0 | 0 | 37 | 0 | 37 |

(*) at rayon boundaries

Southern Tajikistan, the flatlands of Kyrgyzstan, the area around the Karakum Canal and northern Turkmenistan have the same climatic, social and economic potential as Uzbekistan. They also have a high percentage of rural population and surplus of labour force and as a consequence a high poverty level. The highest percentage of rural population is in the country with the least developed fisheries, i.e. in Kyrgyzstan, where the rural population makes 73 percent of the total, and Tajikistan, with 69 percent of the total population (Table 8).

Fish yields in Tudakul and Khauzkhan reservoirs reach 60 kg/ha and 30.9 kg/ha, respectively, while in other reservoirs in the region the yields are only 3–7 kg/ha. The development of fisheries in reservoirs serving irrigation will provide employment and contribute to the diversification of food supply. Development of aquaculture in irrigation systems could also result in a manyfold increase in fish supply to markets.

It is estimated that the fish yield potential of lakes, rivers, and reservoirs in the region is about 100 kg/ha/year. This could provide 200 000 tonnes of fish annually to the markets.

**Table 8**

Distribution of population in the Aral Sea basin (1998) (Statistics Yearbooks 10–20)

| Country | Population | | | | | | |
|---|---|---|---|---|---|---|---|
| | Total | | | Urban | | Rural | |
| | inhabitants | % of total | inhabitants per km$^2$ | inhabitants | % | inhabitants | % |
| Kazakhstan* | 2 710 000 | 6.8 | 7.87 | 1 219 500 | 45 | 1 490 500 | 55 |
| Kyrgyzstan* | 2 540 000 | 6.4 | 19.9 | 685 800 | 27 | 1 854 200 | 73 |
| Tajikistan | 6 066 600 | 15.2 | 42 | 1 880 646 | 31 | 4 185 954 | 69 |
| Turkmenistan | 4 686 800 | 11.8 | 9.7 | 2 109 060 | 45 | 2 577 740 | 55 |
| Uzbekistan | 23 867 400 | 59.8 | 53.2 | 9 308 286 | 39 | 14 559 114 | 61 |
| Aral Sea basin | 39 870 800 | 100 | 25.7 | 15 203 292 | 38.1 | 24 667 508 | 61.9 |

(*) Only provinces in the Aral Sea basin are included

At present there is no regional collaboration for development of fisheries in return waters accumulated in natural depressions. Discharge of drainage water to depressions without outflow has created many lakes and wetlands, the majority of which are shallow. The largest lakes of this type are Sarykamysh and a system of Arnasay lakes. Because of the low discharge capacity of the Syr–Darya River downstream from Chardara storage reservoir (situated at the border between Uzbekistan and Kazakhstan) during the last decade or so water surplus has had to be released into the Arnasay depression during wet years. This usually has happened in winter, as a result of releases from the Naryn–Syr–Darya reservoir cascade. The volume of water in lakes and wetlands formed from drainage water now reaches 40 km$^3$. To ensure that the salinity level is acceptable for fisheries and to protect biodiversity, the water needs to be diluted with freshwater or from time to time flushed out (Fashchevsky, 1996).

The interest of the world in this issue is apparent. Several projects have been partly implemented but discontinued because of lack of funds. These dealt with wetland rehabilitation and drainage water reuse in Uzbekistan (Dengizkul, Kara-Kyr), Turkmenistan (Sarykamysh), Kazakhstan (Kamashbalik, Aktaus, Atsay-Kuvandraya) (Table 9). Other wetland rehabilitation projects are still in progress and are funded by the Global Environmental Facility "Water Resources and Environmental Management in the Aral Sea basin". When completed these projects will assist in the integrated use of irrigation systems as they will contribute to the improvement of hydrological and ecological situation in waterbodies of the Aral Sea basin and to restoration of their biodiversity, including fish production (Ivanov *et al.*, 1996).

While at present no regional networks exist to deal specifically with the use of irrigation systems for fish production, there is a potential for cooperation in the use of irrigation systems for fish production at the regional level. It is believed that the Interstate Coordination Water Commission (ICWC) is a body capable of taking up this function.

**Table 9**
**List of ecologically significant waterbodies and wetlands of Central Asia with transboundary return waters in which maintenance of sustainable regime in the interests of biodiversity is planned for (SIC ICWC data base)**

| No. | River basins and countries | Characteristics of waterbodies including wetlands | | | Flora and fauna |
|---|---|---|---|---|---|
| | | Volume, billion m³ | Actual volume, billion m³ | Salinity g/L | |
| **Amu-Darya basin** | | | | | |
| | Tajikistan | | | | |
| | Turkmenistan | | | | |
| 1. | Lake Sarykamysh | – | 12–15 | 10–12 | Reed, fish, muskrats, birds of passage |
| 2. | Kette-Shor | – | 0.5–0.6 | 4–7 | Reed, fish, muskrats, waterfowls |
| 3. | Lake Raman-kol | – | 0.3–0.4 | 6–7 | Reed, fish, muskrats |
| 4. | Zengi Bobo | – | 0.5–0.6 | 5–7 | Reed, fish, muskrats, etc. |
| 5. | Tedjen | – | 0.3–0.4 | 4–8 | Reed, fish, muskrats, birds of passage |
| 6. | Golden Century Karakum (under construction) | 11–13 | – | 5–8 | Fish, birds of passage |
| | Uzbekistan | | | | |
| 7. | Lake Dengizkol | 5.6 | 2.3 | 7.6 | Reed, water vegetation, muskrat, gulls, egret, ducks and other kinds of birds of passage |
| 8. | Lake Solyonoye (Zimbobo) | 0.150 | 0072 | 3.7 | Reed and other types of vegetation, fish, egrets, ducks |
| 9. | Lake Kara-Kyr | 1.3 | 0.740 | 3.0 | Reed and other types of water vegetation, fish and waterfowls |
| 10. | Ayak-Agitma Fall | 7.3 | 3.34 | 2.5–3.5 | Reed, water vegetation, fish, egrets, ducks, waterfowls |
| 11. | Lake Khadicha | 0.3 | 0.135 | 11.5 | Reed, fish |
| 12. | Lake Mashan-Kol W = 4 000 ha | – | 0.250 | 2-2.5 | Haylands |
| 13. | Lake Akhcha-Kol W = 5 000 ha | – | 0.257–8 | 7–8 | Reed |
| 14. | Lake Kara-Teren W = 2 500 ha | – | 0.150 | 4–6 | Reed, fish |
| 15. | Gulf of Djiltyr-Bas W = 28 500 ha | – | 0.130 | 5–7 | Reed, tugay and other types of water vegetation |

| | | | | | |
|---|---|---|---|---|---|
| 16. | Lake Sudochye<br>W = 58 000 ha | – | 0.880 | 3–3.35 | Various types of water vegetation, fish, muskrats, and birds of passage |
| **Syr-Darya basin** | | | | | |
| | Kyrgyzstan | | | | |
| | Tajikistan | | | | |
| | Kazakhstan | | | | |
| 17. | System of Shieli-Tulekul lakes | | 0.1–0.15 | 4–6 | Reed, fish, waterfowls |
| | Kuan-Darya | N/A | N/A | N/A | N/A |
| 18. | System of lakes in deltaic parts of the river in a quantity of more than 150<br>Djuansadir marshland;<br>Karaterin system, Karakamys, Kartma and Jilakda lakes;<br>Kyzyl Tau lake;<br>Akshi Tau, Kara-Kol, Shamshi Kol lakes;<br>Kara-Kuzek and Kandyozek lakes;<br>Karashalan and Tushi-Bas lakes | N/A | N/A | N/A | N/A |
| 19. | System of Kamyshli-Bas lakes, Raimkol and Tushi-Bas lakes | – | – | – | – |
| 20. | Makpal lake | | | | |
| | Uzbekistan | | | | |
| 21. | Arnasay fall<br>W = 310 000 ha | – | 28–30 | 4–4.3 | |
| 22. | System of Arnasay lakes | – | 0.15–0.2 | 6–7 | Reed, fish and waterfowls |
| 23. | Tuzkane | | 1.5 | 6–9 | Reed, fish and waterfowls |

The top level management organizations from the five countries of the Aral Sea basin are represented by the ministers at the quarterly meetings of the ICWC to discuss the current situation related to water distribution and use, and to formulate water strategy for the forthcoming period. The ICWC consists of three permanent executive bodies: BWOs Amu-Darya and Syr-Darya and SIC ICWC. BWOs Amu-Darya and Syr–Darya are in charge of the operational monitoring of the water limits set by ICWC, and of operation of interstate reservoirs and hydrostructures. The SIC ICWC is in charge of technical policy and manages regional information database on regional water resources.

The ICWC is now also paying attention to the interests of other water resource users including fisheries. It aims at overcoming administrative barriers and tries to involve the general public and private sector, non-governmental organizations and water users in the integrated water resources management both at national and regional levels (Dukhovny and Kindler, 1999).

In our opinion, a developed institutional framework for water management at the regional level and the possibility of regular contacts with governments, related ministries and the general public make ICWC the most suitable structure for information support and development of regional network involving the use of irrigation systems for fish production.

The ICWC Training Centre has branches in all countries of Central Asia. It could train employees of water management operational and maintenance services in fish aquaculture and fishery management of irrigation systems. Assistance of international experts, at least initially, would be essential. This concerns especially the development of small-scale fishery business projects. The ICWC could also assume responsibility for the preparation of strategies and technical support for management of sustainable fisheries in transboundary return waters in the Aral Sea basin. Finally, the SIC, as a permanent executive body of the ICWC, could supervise the implementation of pilot projects testing aquaculture potential of various types of waters supplying and arising from irrigated areas.

## REFERENCES

Anon., 1990. Socio-economic Problems of Aral and Priaralye. Fan, Tashkent. (In Russian)

Anon., 1997. The Main Provisions of Regional Water Strategy in the Aral Sea basin. IFAS, Almaty-Bishkek-Dushanbe-Ashgabad-Tashkent. (In Russian)

Anon., 1998. Biodiversity Conservation. National Strategy and Action Plan. Tashkent.

Anon., 2000. Water Resources, Aral Problem and Environment. Universitet, Tashkent. (In Russian)

Anon., 2001. Rational and Effective Water Use in Central Asia. Diagnostic Report UN SPECA. Tashkent-Bishkek. (In Russian)

Anon., 2001a. Assessment of the Social-economic Damage under the Influence of the Aral Sea Level Lowering. INTAS/RFBR-1733 Project. Final Report. Tashkent.

Dukhovny, V. & Kindler, J. 1999. Developing Regional Collaboration to Manage the Aral Sea basin Water under International and Inter-Sectoral Competition – Water Sector Capacity Building: Concepts and Instruments. A.A. Balkema. Rotterdam/Brookfield.

Fashchevsky, B. 1996. Basics of Ecological Hydrology. Ecoinvest, Minsk. (In Russian)

Ivanov, V. *et al.*, 1996. Review of the scientific and environmental issues of the Aral Sea basin. In: The Aral Sea basin. NATO ASI Series 2., vol. 2: 9–21.

Kamilov, B. The use of irrigation systems for sustainable fish production: Uzbekistan. (This volume)

Zholdasova, M. *et al.*, 1996. Biological basis of fishery development in the waterbodies of the southern Aral Sea region. In: Ecological Research and Monitoring of the Aral Sea deltas: 213–33. UNESCO.